SOURCES AND METHODS IN GEOGRAPHY

Weathering and Erosion

Stephen T. Trudgill PhD
Department of Geography,
University of Sheffield

BUTTERWORTHS
London – Boston
Durban – Singapore – Sydney – Toronto – Wellington

First published 1983

British Library Cataloguing in Publication Data

Trudgill, Stephen T.
Weathering and erosion. — (Sources and methods in geography)
1. Weathering 2. Erosion
I. Title II. Series
551.3'02 QE570

ISBN 0-408-10635-2

Typeset by Scribe Design Ltd, Gillingham, Kent
Printed and bound by Cambridge University Press, Cambridge

FOREWORD

During recent years, geography has been undergoing considerable change. There have been many facets to this change, but one underlying theme is the adoption of a more rigorous approach to geographical enquiry, wherever this is appropriate. It has been reflected in numerous ways: in the greater emphasis which is placed upon quantitative and statistical methods of data collection and handling; in the attention given to the study of process as opposed to the description of form in human as well as in physical geography; and in the use of an inductive rather than deductive philosophy of learning.

What this means in practical terms is that the student and teacher of geography need to be acquainted with a wide range of methods. The student, both at school and in higher education, is increasingly becoming involved in projects or classwork which include some form of individual and original research. To be equipped for this type of study he/she needs to be aware of the sources from which he/she can obtain data, the techniques he/she can use to collect this information and the approaches he/she can take to analyse it. The teacher similarly requires a pool of empirical material on which he/she can draw as a source of class exercises. Both must be able to tackle geographical problems in a logical and scientific fashion, to construct appropriate explanatory hypotheses, and test these hypotheses in an objective and rational manner.

The aims of this series of books are therefore to introduce a range of sources which provide information for project and classwork, and to outline some of the methods by which this material can be analysed. The main concern is with relatively simple approaches rather than more sophisticated methods.

The reader will be expected to have a basic grounding in geography, and in some of the books a working knowledge of mathematical methods is useful. The level of detail and exposition, however, is intended to make it possible for the student, with little further reading, to gain a basic understanding of the selected themes. Consequently the series will be of particular interest and use to students and teachers involved in courses in which practical and project work figure as major components. At the same time students in higher education will find the books an invaluable guide to geographical methods.

M.A. Morgan D.J. Briggs

PREFACE

The aim of this book is to provide an introduction to project work and a data source in the fields of weathering and erosion. Many school teachers with whom I have had contact have experienced a need for simple and effective practical methods and materials which, as well as providing a vehicle for the teaching of principles and processes, are also cheap and viable. Therefore I have attempted to outline techniques that inevitably involve assorted tin cans and polythene bags, simple chemical tests and experimental work that a basically equipped laboratory can handle. The approach is therefore selective, but reference is made to more sophisticated techniques which may be followed up via the reading material included for more detailed university-level project work. Inevitably some techniques are over-simplistic, especially if applied unthinkingly to complex field situations, but if care is taken in the selection of material, useful experimental work can be undertaken in the laboratory and in the field. This is especially true if the measurement technique is not regarded as simply an end in itself but as a means of gaining information to help to understand the operation of the real world. Experiments and observations should be organised so that they test or observe environmental processes rather than just being used as techniques.

Weathering and erosion are fundamental to the topic of geomorphology, that is to the understanding of the evolution of landforms. In addition, there are wide applications in the fields of plant nutrition, water quality, agriculture and engineering. The book is therefore intended not only for sixth-form geographers, and as an introductory text for polytechnic and university geographers, but also for students at many levels of geology, environmental science, biology and agriculture. It is regarded as a companion volume to the earlier books in this series and, while there is some overlap, reference is often made to statistical tests and other material covered in the previous books and which are not duplicated here. It is not a text book of weathering and erosion processes – adequate geomorphological texts exist for this purpose and these are included in the section on further reading. It is intended that the outlines given in this book will provide a working basis for project work. The data sets are intended to provide a basis for class interpretation.

I have always been impressed by the value of practical work in that it gives a sudden grasp of meaning and an upsurge of confidence in students who have mastered a technique and gained interpretable results; it is hoped that this book may be a positive contribution towards this learning process.

S.T.T.
Harthill, 17.1.81

ACKNOWLEDGEMENTS

I am indebted to my Bristol PhD supervisor, Dingle Smith, whose enthusiasm for field research and field measurement inspired much of the approach adopted in this book. Dave Briggs has provided much encouragement and useful criticism at the so-many stages of this book. Staff of the Field Studies Council, notably Tony Thomas and Maggie Calloway, have provided constant stimulation for the cause of viable field techniques and teachers attending FSC courses I have run have constantly demanded simple, viable teaching techniques! I would especially like to thank the friends and colleagues with whom I have the privilege of working and who keep me humble, sane and supplied with tea and coffee (or drink mine), particularly Bob Crabtree, Adrian Pickles, Keith Smettem, Pete Walker, Nigel Coles and Tim Burt. Many thanks also to my parents for support, to Paul Coles and the other Geography Department cartographers and the typists who wrestle with my atrocious drafts and manuscripts.

CONTENTS

ILLUSTRATIONS

TABLES

PART I

INTRODUCTION

CHAPTER 1 INTRODUCTION

1.1 THE NATURE OF WEATHERING AND EROSION

If the stones of buildings are observed it can often be seen that, unless they have been cleaned regularly, the older the buildings are the more discoloured they have become. In towns and cities this is partly due to discoloration with soot and to chemical reactions with acid gases from industrial and domestic air pollution. Even in country areas, however, well away from industry, the alteration of building stone may be seen as time progresses. The stone is exposed to rain water, oxygen in the air and changes in temperature. Thus it may become dissolved and oxidised, or stressed by expansion and contraction during heating and cooling. These processes may be collectively termed ***weathering*** in its original sense: ***the exposure of rocks to the weather*** and the alteration of rock by rain, air, heat and frost. Weathering can be more broadly defined as the chemical alteration and physical breakdown of rock material in response to environmental conditions. Weathering is usually thought of as operating ***in situ*** and can thus be differentiated from ***erosion*** which is the loss of material from its pre-existing position and its transport to other locations. Erosion may involve the movement of unaltered rock material or the removal of weathering products (***Figure 1.1***).

Weathering and erosion are important in a wide variety of contexts. They act together to shape the nature of the surface of the earth and are thus fundamental to the study of geomorphology. Soil formation and development are often dependent on weathering reactions, not only in terms of the weathering of soil parent material to form soil but also in terms of the weathering of minerals present in the soil itself. Soil erosion may affect the course of soil development and the agricultural use of a soil. Soil weathering reactions provide many of the mineral elements necessary for plant growth and are thus important to agriculture and plant ecology. The products of weathering and erosion may be transported by surface runoff waters, and thus the nature of weathering and erosion in drainage basins is important in fluvial and hydrological contexts as an influence upon water quality.

It can therefore be seen that while the most obvious signs of weathering may involve, to the casual observer, little more than the alteration of building stones, it is a process

Figure 1.1 Weathering and erosion of an old wall composed of coarse-grained, loosely cemented calcareous sandstone. The wall has been re-cemented between the blocks but over 200 years the sandstone has eroded in relation to differential cementation of the grains (Photograph: S.T. Trudgill)

of far-reaching importance and one which involves many reactions such as those hidden within the soil. These are less obvious and may be more difficult to study.

Similarly, the spectacular erosion of cliffs by the sea, the loss of river banks and landslips during floods or the blowing of soil across roads during gales may make the news headlines but it is often the more subtle processes, involving the erosion of minute amounts of material, which are more widespread and which may have a greater effect in the long term.

1.2 THE STUDY OF WEATHERING AND EROSION

Clearly, weathering and erosion are important topics which may affect human society. This effect may be direct, through the spectacular events mentioned above or through systems upon which society depends such as agricultural systems or water-supply systems. Alternatively, the effects may be less direct through involvement in fundamental environmental processes such as land-form evolution or soil development. The less direct effects are no less relevant to society than the direct effects; simply it takes longer for the effects to be seen and therefore the concern is less immediate.

It is also clear that some effects are far more difficult to study than others. Spectacular erosion events can be fairly readily documented. For example, cliff erosion can be studied by comparing old maps with present-day maps and the individual cliff-fall events are often recorded in newspapers. However, the more subtle processes may operate only slowly and their effects may not be readily discernible except over several years' observation. In this case any short-term measurements have to be extremely accurate. Moreover, some of the processes can only be tackled indirectly; for example, in the case of weathering reactions in the soil, by a study of weathering products present in runoff waters rather than necessarily by a study of the actual processes operating within the soil itself.

It is because of the problems of slow rates of reactions and of the inability to measure some processes directly that some laboratory experimentation may often be useful. In the laboratory, rates of reaction can be speeded up and conditions can be controlled. Thus reactions may be readily observed and the importance of various factors can be evaluated by the use of control experiments. However, this approach, in itself, leads to a further problem – that of making the laboratory work relevant to the field situation.

Field processes are often complicated and involve the simultaneous actions of many chemical, physical and biological processes in varying proportions over long periods of time. Simple and rapid laboratory tests under controlled conditions are useful to demonstrate how processes operate individually and can be made very precise, but they do not necessarily elucidate how processes operate together in nature. Nevertheless, control experimentation in the laboratory is usually an important first step in unravelling complex natural processes and in specifying the role of individual processes in producing an overall result.

Field experimentation does, however, provide an important approach. Here, the processes can be studied while they operate naturally. Although this is true, the problem is that it is often very difficult to control and separate the various factors which may be involved. The results of field experiments may thus be ambiguous and difficult to interpret unless the experiment is designed carefully. In addition, field instruments have to be more robust than laboratory instruments in order to survive in the open. This may make them less precise, which may be a problem if the processes are operating very slowly. The challenge here is thus to undertake field experiments in which factors can be isolated and identified and where instrumentation can be used which is both robust and accurate. Lastly, it should be added that it is often difficult to study the operation of natural processes in the field without disturbing them in some way by the introduction of a measurement or sampling device. Nevertheless, field experimentation and measurement may often yield important information on the operation of natural processes and is, in many ways, the most desirable approach to adopt.

1.3 RESEARCH DESIGN

There are therefore many problems involved in the study of weathering and erosion and if experiments are to be successful a number of limitations must be borne in mind. One of these limitations lies in the useful differentiation between ***demonstrating*** (or illustrating). and ***testing*** (***Figure 1.2***). The distinction is between (1) demonstrating how a process might work on the basis of a few experiments or observations and (2) the statistical testing of the universality of the operation of a process, based on a large number of samples. For example, two observations on the calcium content of a soil, one at the top of a slope and one at the bottom of a slope may ***illustrate*** the effects of leaching downslope. Thirty samples from a range of slope tops, however, compared statistically with thirty from a slope-foot site would give a more ***rigorous test*** of the differences between the sites of and the operation of the process. Often, in a teaching situation, it may only be possible, because of time limitations, to take a few observations. In this case, the study sites will have to be carefully selected in order to minimise the effects of other factors which could alter the picture but which are not being measured. Thus if soil leaching is to be studied it would be necessary to compare sites with similar bedrock, similar soil parent material and similar vegetation. In practice, this is often difficult since

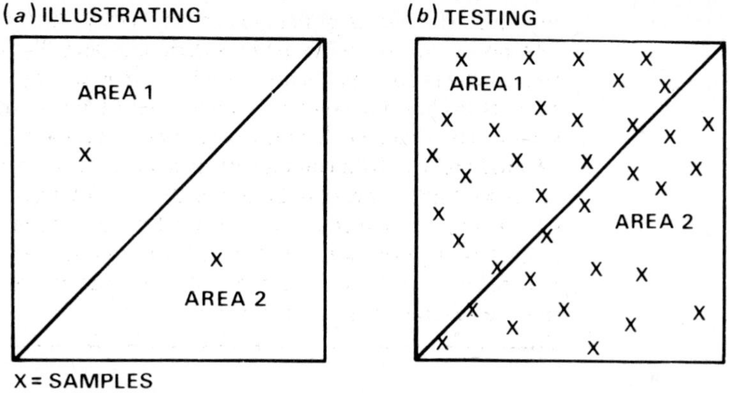

(*a*) ILLUSTRATING

(*b*) TESTING

X = SAMPLES

Figure 1.2 Sampling for illustrative purposes and for the purpose of statistical testing

these factors may tend to vary together, but an effort must be made to do this otherwise the results will be uninterpretable in any simple fashion. Another alternative is to study the co-variation of the many factors, but this would require a large sampling programme. Frequently only a transect is possible which, when drawn up, has specific value in that it ***demonstrates*** the points being made (***Figure 1.3***) but which is of ***limited statistical validity.***

In making the distinction between demonstrating and testing there is no necessary sense of advocacy that one is superior to the other; simply each set of observations will

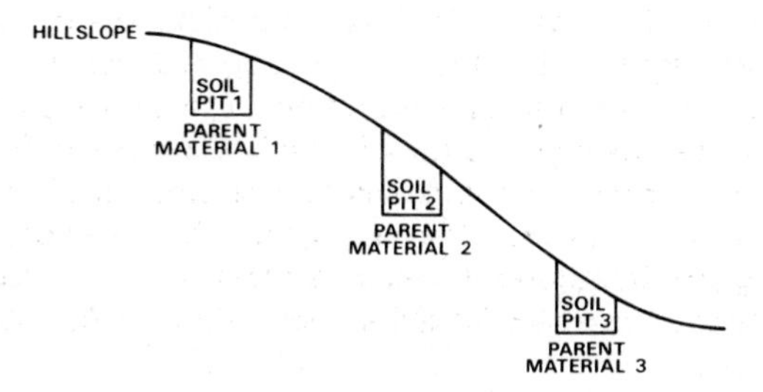

Figure 1.3 Layout of illustrative transect of downslope soil changes

have its limitations and therefore any project will need to have its objectives clearly thought out and specified beforehand.

Laboratory tests are particularly useful in demonstrating the operation of processes, particularly when only a piece of rock and a few chemicals or simple apparatus are all that is required. A simple approach may be unsophisticated but requires little capital equipment and little space and does not entail a field trip. Classroom and laboratory tests are thus included in this book as well as field work.

Field observations can obviously be made at a number of levels of complexity. A few observations at carefully selected sites can often demonstrate the operation of major processes, as discussed above. Using what may be termed an 'outdoor laboratory' approach, it is often possible to select sites where as many factors as possible are constant while one or two vary. For example, it is possible to study the erosion of different rocks in similar environments or the rate of leaching in different soils formed on contrasting rock types.

It is only at a higher level of sophistication that more rigorous work can be attempted. This work is often also the most repetitive. Fifty observations of the solute load of stream waters draining from rock type A and 50 from rock type B are time-consuming but necessary if one is to go beyond a relatively simple illustrative exercise. Careful thought is also required. Field measurements can far too easily be taken without enough planning of what they might test. It is important to try to test an hypothesis and to be sure that the hypothesis is firmly bedded in a wider body of theory. It is also important to ensure that as many factors as possible are kept constant while those of interest vary, that the techniques used are actually testing the hypothesis in question, and thus that the results are as unambiguous as possible. This is difficult but it is important not just to test site A against site B to see if there is a difference. This is too casual and if there is (or if there is not) a difference then the results may well be uninterpretable, or at least have several possible alternative explanations. Such an approach to gathering information is only appropriate if little or nothing is known about an area or phenomenon (hypotheses for testing can then be formulated from the facts collected). Instead, because a knowledge of process forms a basis for scientific explanation, it is important to specify why there should be a difference between any sites chosen for study and to make sure that the samples are taken so that the reason hypothesised is, in fact, tested. It is also important to understand causal mechanisms involved in processes and not merely to correlate observations or to show statistical differences between data sets. In other words it is important to be able to explain any results. This approach is not without its problems, largely because of the complexities which may be involved in nature. Therefore, inevitably, the simple experiments and observations which illustrate processes are often those described in this book. It is, however, necessary to make these qualifications so that the limitations of such approaches are known.

Whatever the experiment or measurement, be it a field or laboratory one, it is useful to plan it and write it up under the following headings:

(1) ***Aims.*** Give the reasons for the project and state what hypothesis is being tested or what it is intended to show, with reference to the general body of theory involved.
(2) ***Methods.*** If a method is copied from a text it is usually sufficient to refer to the standard procedure. However, any deviations from this should be recorded and non-standard or new procedures should always be described.
(3) ***Results.*** All results should be carefully recorded and presented in the form of tables or graphs as appropriate.
(4) ***General discussion.*** Discuss the results in relation to the aims and the theory being tested and also in relation to any other similar projects being carried out. Also discuss the results in relation to any problems encountered during the project and any errors which might have occurred, and also suggest how problems may be overcome and how future work might usefully be carried out.

1.4 APPROACHES TO THE STUDY OF WEATHERING AND EROSION

Weathering and erosion systems can be studied at several points. These are as follows (*Figure 1.4*):

(1) the potential for weathering and erosion inherent in the environment;
(2) the susceptibility of the material involved to the potential existing in the environment;

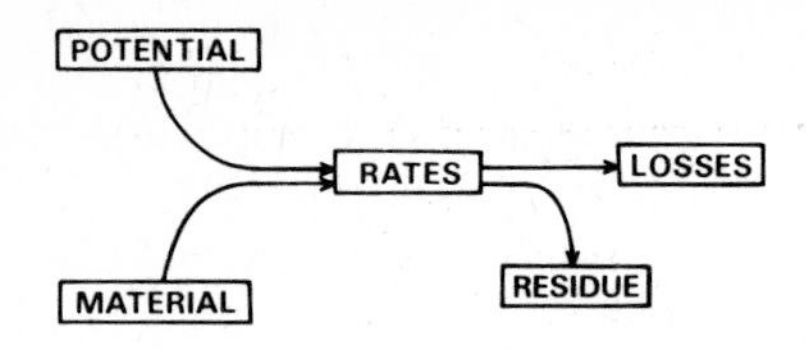

Figure 1.4 Points at which weathering and erosion can be studied

(3) the rates of operation of processes at the site of reaction;
(4) the rates and amounts of losses of material from the site of reaction;
(5) the residual material left behind after reaction.

Although the materials and processes involved are different in weathering and in erosion, the same general principles apply to both.

For example, in chemical weathering, the acidity and the oxidation potential of an environment may be studied. These may be studied in rain water, soils or stream water. The susceptibility of rocks and minerals to solution by acids or to oxidation can be studied. Subsequently, the losses of chemical elements from the mineral may be observed as well as the chemical nature of the weathering residue.

In physical weathering, a study of temperature changes can, for example, reveal the potential for freeze–thaw action. The susceptibility of material to this type of action, the losses of material and the nature of any material remaining ***in situ*** after the operation of the process can all be evaluated.

In erosion, the forces involved in the impact of raindrops, waves or river flow can be studied. Then the susceptibility of rock and soil materials to these forces can be analysed, the erosion mechanisms evaluated and the rates of loss of sediments and rock surface material measured. A study of any material remaining after erosion will help to reveal which portions of the material are most resistant to erosion.

It is with the study of the factors such as those outlined above that this book is concerned. Although it may be difficult to separate the various processes precisely at the relatively simple level of theory and technique at which this book is written, the recognition of the points identified in ***Figure 1.4*** will be useful.

PART II

WEATHERING

CHAPTER 2 MECHANICAL WEATHERING

2.1 INTRODUCTION

Mechanical weathering involves the breakdown of large masses of solid rock into smaller fragments. It provides material which may constitute soil parent material or sediments and it also provides fresh rock surfaces on which chemical and biological weathering agencies can act. Surface area is an important factor in chemical weathering and processes such as solution occur more rapidly as surface area increases. Thus, mechanical action may have a strong influence upon chemical processes.

Mechanical weathering occurs when the rock is stressed in some way, the rock breaking up under stress along lines of weakness. Rock fragments may be produced or the rocks may be split into individual minerals or crystals along crystal cleavage planes (*Figure 2.1*).

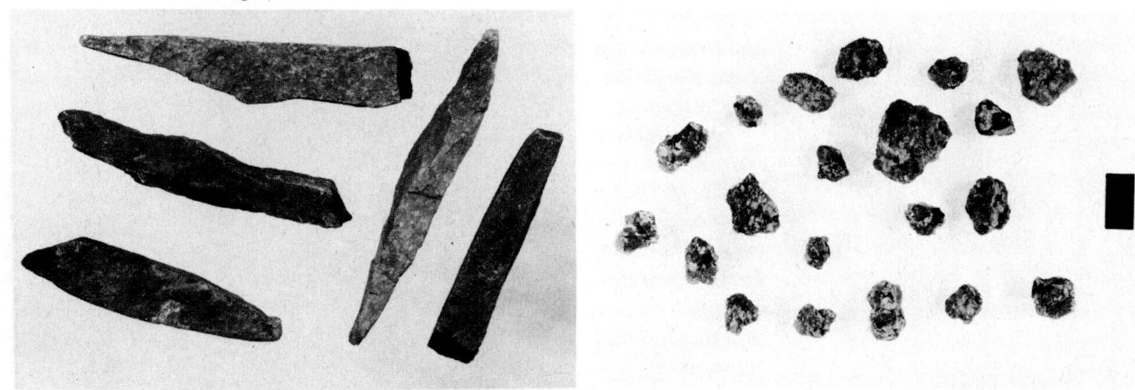

Figure 2.1 (a) Mechanical weathering of slate showing break-up along lines of weakness to produce pencil-like forms. Ordovician slate from mountain top in N. Wales (Photograph: J. Owen) (b) Mechanical weathering of granite from the summit of Cairn Gorm, Scotland, showing the effects of frost action in breaking up the rock along crystal faces. Scale bar = 1 cm (Photograph: J. Owen)

Rocks may be stressed by the uptake of water and a consequent expansion. Temperature change can also be important and excessive heating or cooling can cause volume changes which stress the rock. In addition, the growth of salt crystals can result in fracture. In all cases, even where freezing or heating are the main agents of stress, water uptake is involved to a greater or lesser extent. In the following sections processes of water uptake and, subsequently, salt crystallisation will be considered.

2.2 WATER UPTAKE

One of the main controls on the ability of a rock to absorb water is the porosity. Rocks are composed of crystals or mineral grains, held together by interlocking or by chemical cements. The crystals or grains rarely fit perfectly together, however, and small holes or pores occur between them. It is there that the water enters when the rock is wetted. Porosity is a measure of the volume of the pores, and thus is also a measure of the ability of the rock to take up water.

In order to study the porosity of a rock, several samples may be immersed in water and mass increase can be used as in indication of water uptake. Rocks of similar size and shape should be used in order that the different rock types may be compared in a standard form.

As an initial hypothesis it may be suggested that sedimentary rocks may be more porous than metamorphic and igneous rocks. This is logical since metamorphic rocks have usually been compressed, thus reducing pore space. Igneous rocks will have formed by the growth of interlocking crystals and thus pore space will be low. With sedimentary rocks, however, their granular nature may mean that there is a high inter-granular pore space. Grain size and cementation will be important controls on porosity. Small, regular grains will pack more tightly, leaving less pore space between the grains than occurs with large, irregular grains. A cement may be a simple point cement, that is, one which occurs at the point of grain contact, or it may fill the spaces between the grains, acting to minimise pore space. It will be useful to study each rock with a hand lens (or in thin section under a microscope) before using the techniques outlined below for studying water uptake. Suggestions can be made as to the order of expected water uptake based on observations of visible porosity and cementation.

2.2.1 Method

Equipment

a set of glass beakers large enough to hold sufficient water to cover the rock samples by about 5 cm depth
weighing balance
drying oven
desiccator with silica gel (or calcium chloride) able to absorb water

Samples

a set of identified rock samples

Procedure

(1) Chip the rock samples to approximately the same size and shape as each other using a geological hammer or a hammer and chisel*, taking care to protect the eyes from splinters by wearing goggles. Samples of about 2 cm^3 are recommended. If a rock cutter is available then cubes of standard size can be cut precisely.
(2) Rinse and then dry the rock samples in an oven at 105 °C for 24 h to vaporise any moisture from the pore spaces.
(3) Place the sample in a desiccator for 1 h or until at room temperature.
(4) Weigh the dried, cooled sample of rock†.
(5) Immerse each sample in water, each in a separate container, for 1 h‡.
(6) Remove the sample and touch the surface with blotting paper to remove any surplus surface water.
(7) Reweigh the sample.

The technique is further discussed by Cooke (1979).

* Do not use geology hammers on chisels as some makes of hammer splinter when used in this way; use a lump hammer.
† In some cases rock samples may already be dry, in which case steps (2) and (3) may be omitted.
‡ The rocks may be immersed for 10 h or 24 h, as convenient, but most rocks of the small size recommended will have absorbed 80–90% of their capacity within 1 h.

2.2.2 Results

Mass increases can be listed in order and the uptake can be expressed as a percentage of the original mass, i.e.:

$$\% \text{ water uptake} = \frac{m_2 - m_1}{m_1} \times 100$$

where m_1 = initial dry mass, and m_2 = wet mass after immersion. The data can then be listed in order of percentage mass increase.

Any hypothesis that was suggested at the outset can be tested using statistical methods. For example, if ten samples of sedimentary rocks and ten of igneous rocks were used then a statistical test could be used to ascertain whether there is a significant difference between the two rock types.

Simple illustrative test

Percentage water uptake (1 sample of each rock)

Chalk	12
Sandstone	2
Granite	0.1

Sedimentary rocks (% water uptake)	Igneous rocks (% water uptake)
11	0.1
10	0.09
9	0.1
0.7	0.05
0.9	3
0.7	8
5	0.09
8	0.07
9	2
0.9	0.08

Statistical test (see table on left)

Hypothesis: examination with a hand lens suggested that several samples of sedimentary rocks were more porous than samples of igneous rocks; therefore it is suggested that the former will have a greater uptake of water.

Visual interpretation of these data suggests that the data sets are almost mutually exclusive but that there is enough overlap to make a statistical test useful. These tests are discussed in Briggs (1977a) (the first book in this series) on pages 36–42 (Chi-square), 122–128 (Mann–Whitney), 87–94 (Spearman's rank) and 91–93 (student's 't'). For example, a Mann–Whitney test can be performed on the data above to show that there is a significant difference between the two data sets, thus supporting the initial hypothesis.

2.2.3 Discussion

The interpretation of the results should include a discussion in the light of the variables influencing porosity (such as cementation, grain size, degree of metamorphosis, crystal interlocking) and also a prediction of the stability of the rocks under mechanical weathering stresses. The rocks which take up most water should be the most prone to mechanical disintegration by temperature changes and salt crystal growth. It may also be necessary to discuss the possibility of the presence of anhydrous minerals, which may take up water directly into the mineral structures as well as into the pore spaces (***see*** section 2.3).

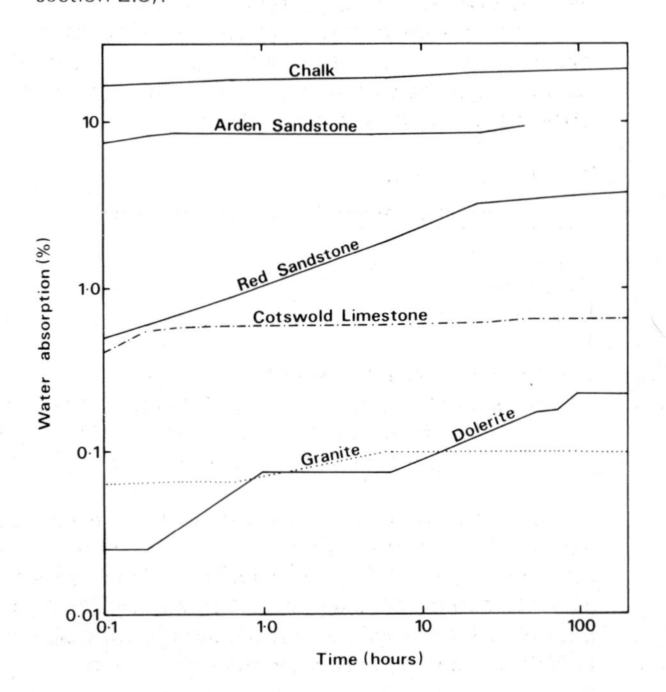

Figure 2.2 Water uptake by rocks of differing porosity; absorption is indicated by percentage increase in mass (From Goudie, Cooke and Evans, 1970)

2.2.4 Case study: Water uptake by eight rock samples (Goudie, Cooke and Evans, 1970)

Eight rock samples of approximately the same size were weighed and immersed in water and reweighed at repeated intervals. The results are shown in ***Figure 2.2.*** It can be seen that water uptake is either rapid and high (e.g. chalk) or slow and low (e.g. dolerite). Chalk has a large volume of inter-granular pore spaces of 0.001–0.1 mm in size comprising as much as 10–20% of the volume of the rock, though this is much smaller when the rock has been compressed. The sandstones and Cotswold Limestone often have point cements (***Figure 2.3***). The cement may be siliceous (in the sandstones) or calcareous (in both sandstones and limestones). Rocks with a lower water uptake have only hairline inter-crystalline cracks, such as those shown for granite in ***Figure 2.4.***

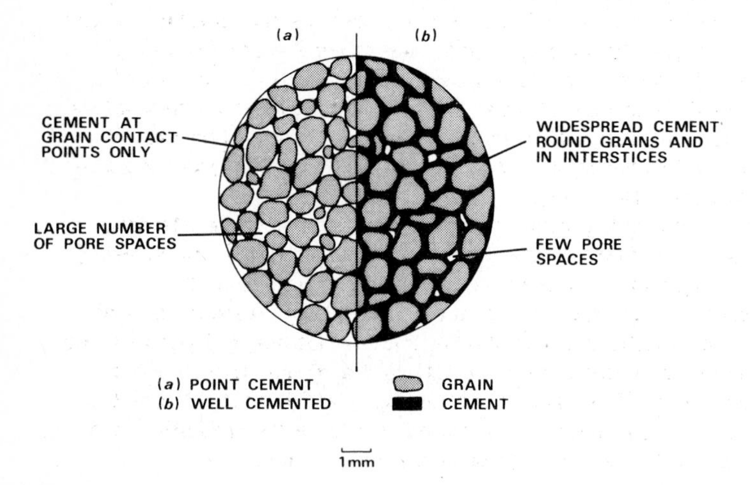

Figure 2.3 Point cement in a sedimentary rock (sandstone). Note the occurrence of cement at the points of contact between the grains and the high inter-granular porosity

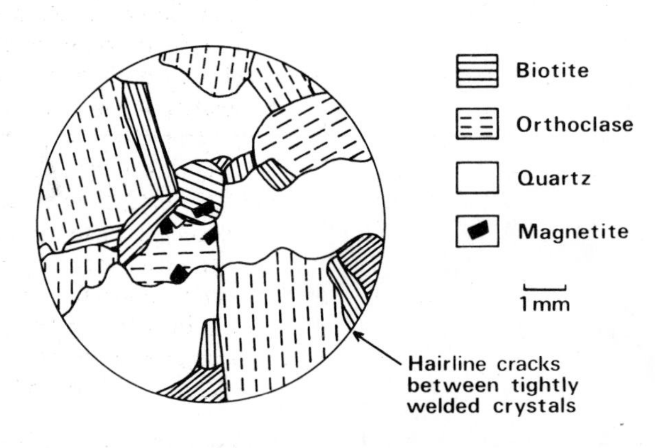

Figure 2.4 Thin section of a granite rock showing the scarcity of pore space, except for narrow inter-crystalline cracks

2.3. WETTING AND DRYING

Alternate wetting and drying may introduce stresses into a rock. Firstly, the minerals may take up water and expand. This uptake of water is termed hydration and water actually becomes part of the crystal lattice of the mineral. Hydration of minerals can cause considerable stress and is thought by some research workers to be a principal cause of the mechanical disintegration of rocks, even in situations where freeze–thaw is traditionally thought to be the main process operating (White, 1976). Indeed, it is likely that water uptake in itself does not cause many stresses in the rock unless some reaction such as hydration occurs (or unless there is salt crystal growth, ***see*** section 2.5). With alternate hydration and dehydration there will be alternate expansion and contraction. This will result in the weakening of crystals. Eventually, the minerals will tend to break along lines of weakness, usually along crystal cleavage planes.

The uptake of water does not necessarily occur in nature as a result of total immersion but it may take place from water vapour present in a moist atmosphere or from surface dew. However, it is usually far easier to experiment with total immersion, unless a humidity chamber is available (this is a sealed cabinet where the moisture-vapour content can be controlled and measured; such a chamber is described and was used by Kwaad (1970), but warm, moist conditions can be crudely simulated by using a drying oven at a low setting with a dish of water placed in it).

As well as water absorption by the crystal lattice, water may also become adsorbed onto the surface of clays, and, in addition, chemical reactions may occur – for example, iron oxides may be transformed into iron hydroxides. Swelling may frequently accompany these processes. It can thus be noted that there is no clear division between many chemical and mechanical actions. It can also be seen that not only will porosity and water uptake be involved in rock disintegration by wetting and drying, as discussed in section 2.2, but also the presence of clays and other minerals capable of being hydrated. Thus, the study of the rock and the suggestion of hypotheses about the susceptibility of rocks to wetting and drying will have to take these factors into account.

2.3.1 Method

The technique is the same as in 2.2 except that after immersion the sample is dried at 60–80 °C for at least 24 h to drive off all the water. Then, the cycle of wetting and drying is repeated several times and any changes in mass are recorded (after cooling in a desiccator). In addition the sample can be inspected for signs of fragmentation. It should be noted that drying is achieved by heating which itself causes stress. Air drying for 2–3 days in a warm, dry place could therefore be used as an alternative to the use of an oven at 105 °C.

2.3.2 Results and discussion

Few rocks will disintegrate by this process alone unless the wetting and drying cycles are repeated at least 10–20 times or unless the rock contains a mineral which is readily hydrated such as anhydrite.

2.4 HEATING AND COOLING

The technique to assess the effect of heating and cooling is the same as in section 2.3 except that the sample is not immersed in water; heating can take place up to 105 °C or above.

Well-known case studies of these two experiments are the works of Blackwelder (1933) and Griggs (1936) which established that heating and cooling were relatively ineffective by themselves (Ollier, 1969, p. 15). Under the experimental conditions used in their studies the presence of water caused far more rapid disintegration of the rock samples used than was the case where water was absent. Although experiments with small rock samples cannot reproduce the fatigue and strains produced in large rock masses over time in nature, they can be useful to ascertain the relative susceptibility of individual rock types to weathering under experimental conditions.

A quantitative study of rock-surface temperature changes has been undertaken by Smith (1977) in a limestone valley in the North-west Sahara. Temperature ranges for two positions, (a) west facing and (b) east facing, are shown in ***Table 2.1***. Near the Equator the sun is largely overhead but west-facing slopes generally experience the greater temperature range. Smith concluded that sites with the greatest temperature range ***and*** moisture range would be the most prone to weathering.

TABLE 2.1 DIURNAL TEMPERATURE RANGES RECORDED ON ROCK SURFACES IN N.W. SAHARA (AFTER SMITH, 1977)

Site	Depth (cm)	Maximum temperature (°C)	Minimum temperature (°C)	Maximum range (°C)
a	0	9.1–19.1	1.5–3.9	5.2–17.5
	– 5	7.9–22.2	3.1–5.1	2.8–19.1
	–10	7.9–21.0	3.5–5.3	2.6–16.5
b	0	8.1–18.9	3.7–4.2	3.9–15.2
	– 5	8.0–17.5	4.2–5.2	2.8–13.3
	–10	7.9–15.1	4.3–5.6	2.3–11.8

2.5 SALT WEATHERING

Salt weathering can be an important process of rock disintegration in semi-arid climates where high rates of water evaporation act to concentrate dissolved salts. As evaporation proceeds, salt crystals begin to grow in the pore spaces in the rock. If the force exerted by the growing crystal is greater than the mechanical strength of the rock then salt crystal growth can act to prise the rock constituents apart. Thus, in a weakly cemented sandstone, salt crystals may grow in the pore spaces in the rock, either forcing the individual grains apart or forcing flakes of rock away from the main body of the rock. This process can be particularly effective in semi-arid regions and, as well as weathering natural rock faces, it constitutes a major hazard in the disintegration of building stones in these areas. The process can also operate along the sea coast where salt spray is blown onto sea cliffs and where the water subsequently evaporates.

It has been questioned whether the force of salt crystal growth is sufficient to prise rocks apart. The point can be made that if salt water enters the rock then salt crystal growth could take place outwards in the opposite direction to which the water entered, thus taking the path of least resistance, rather than growing into the rock mass where there is some resistance. In reply to this argument it has been suggested that crystal growth could occur at the surface first, thus trapping solutions inside the rock. Here evaporation could not occur, but if the solution was chemically saturated with the salt then crystal growth could occur giving rise to pressure on the rock material. However, what is clear is that the saline solution has to penetrate into the rock mass before disintegration occurs; simple wetting of the surface is relatively ineffective. Rock porosity is thus, once again, a factor of considerable importance, as suggested in section 2.2.

2.5.1 Method

The same type of equipment, samples and procedure as for 2.2 and 2.3 are needed except that when wetting occurs, saline water is used instead of distilled water. The simplest technique is to add 10 g of sodium chloride to each wetted sample before heating. As the water evaporates salt crystals will readily grow on the surface but it will usually take several cycles of saline wetting and drying before the solution permeates into the rock and disintegration starts.

2.5.2 Results and discussion

Results can be most easily plotted in terms of the number of fragments into which the rock disintegrates. In terms of mass loss a problem exists in that salt will permeate into the rock and a mass increase due to salt uptake will be recorded initially, before a subsequent mass loss is recorded as fragments begin to break off. If removal of the salt is desired, it will be necessary to soak the samples in distilled water in order to dissolve out all the salt before drying, cooling and weighing. However, in practice it is often difficult to leach out all the salt present in the rock pores and so it is better to proceed with cycles until large fragments have fallen off the rock; then the salt mass error will be small relative to the rock fragment mass loss. Alternatively, a simple count of fragments often suffices.

Table 2.2 shows data for a simple salt-weathering experiment and ***Figure 2.5*** shows examples of Carboniferous Limestone, Recent carbonate rocks and schist after five salt weathering cycles. While the Recent carbonate rock is the most porous (about 10%), and the Carboniferous Limestone and schist have a porosity of about 0.01%, it is the schist which has broken down most readily. This is because it has many cleavage planes along which disintegration occurs. The most porous rock has, in fact, not broken down as the pores are very large and there is ample space for salt crystals to grow without any pressure being exerted on the rock; also it is very strongly cemented. Thus not only overall porosity but also pore size and strength of the rock can be important.

TABLE 2.2 MASS LOSS (g) OF SANDSTONE AND LIMESTONE TABLETS AFTER WETTING AND DRYING CYCLES WITH AND WITHOUT NaCl

Rock	Distilled water	Salt solution
Devonian Sandstone		
1	0.0070	0.0288
2	0.0101	0.0074
	Mean 0.00855	Mean 0.0181
Carboniferous Limestone		
1	0.0034	0.0129
2	0.0025	0.0389
3	0.0040	0.0865
4	0.0016	0.1887
5	0.0048	0.0169
6	0.0054	0.0078
	Mean 0.0036	Mean 0.0191
Weathered limestone		
1	0.0057	0.0163
2	0.0057	0.0163
3	0.0059	0.0112
4	0.0066	0.0075
5	0.0053	0.0140
	Mean 0.00587	Mean 0.0123

Figure 2.5 The effects of salt weathering on blocks of Carboniferous Limestone (left), Recent porous carbonate rock (centre) and schist (Photograph: J. Owen)

2.5.3 Case study

Work by Goudie ***et al.*** (1970) compared the effectiveness of different salts in causing the disintegration of one rock type (***Figure 2.6***). Sodium sulphate was found to be the most effective after 40 cycles (***Table 2.3***) and this conclusion was also supported by Cooke (1979). Sodium sulphate has a relatively large crystal size which may account for its effectiveness.

In addition, Goudie ***et al.*** used sodium sulphate to compare different rock types (***Figure 2.7*** and ***Table 2.4***). The data shown here can be compared with the data in section 2.2. The similarity between the water uptake (***Figure 2.2***) and salt weathering (***Figure 2.7***) can be clearly seen. The samples with rapid water uptake (chalk, Arden Sandstone) disintegrate quickly while those with low water uptake (e.g. dolerite) break down more slowly. However, the relation is not a simple one and, as well as porosity, the strength of cementation is also involved. Thus, although the Cotswolds Limestone had a lower water uptake than the Arden Sandstone, its disintegration is the same because of

Figure 2.7 Effect of sodium sulphate – the salt weathering of different rocks, 40 salt weathering cycles (After Goudie, Cooke and Evans, 1970)

Figure 2.6 Effectiveness of salt type during salt weathering cycles (After Goudie, Cooke and Evans, 1970)

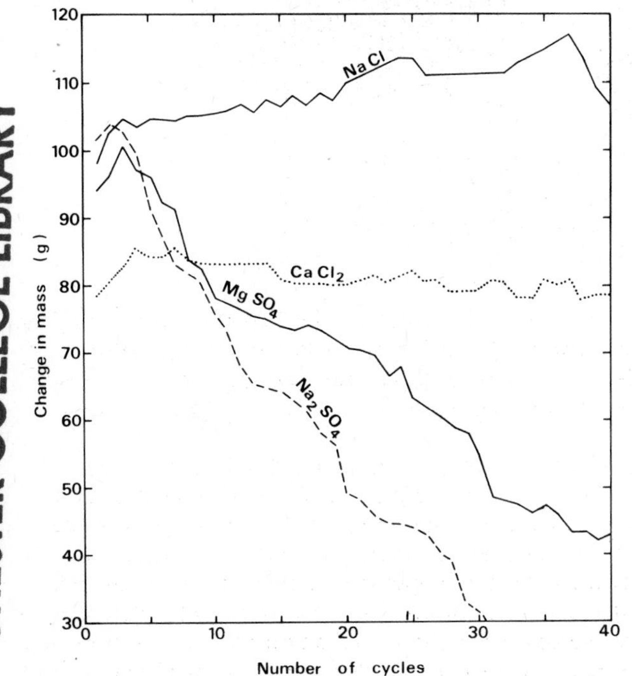

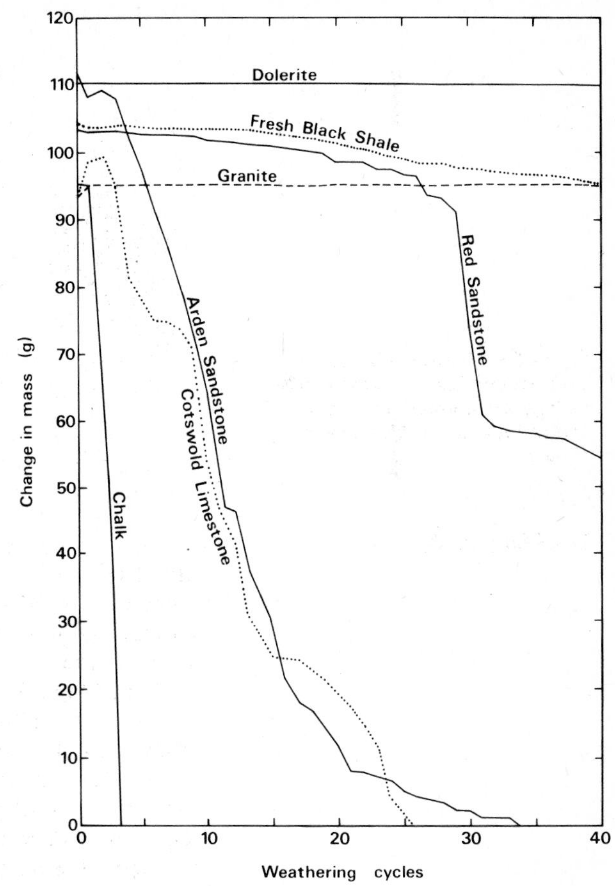

TABLE 2.3 SPLITTING OF ARDEN SANDSTONE (INTO PARTICLES OVER 1 g) BY DIFFERENT SALTS (AFTER GOUDIE, COOKE AND EVANS, 1970)

Salt	Sample	Number of splits
$CaCl_2$	1	1
	2	0
$MgSO_4$	1	1
Na_2SO_4	1	4
	2	5

TABLE 2.4 SPLITTING OF SEVERAL ROCKS (INTO PARTICLES OVER 1 g) BY Na_2SO_4 (AFTER GOUDIE, COOKE AND EVANS, 1970)

Rock type	Sample	Number of splits
Chalk	1	1
	2	2
Red Sandstone	1	0
	2	2
Arden Sandstone	1	5
	2	6
Cotswold Limestone	1	3
	2	1

TABLE 2.5 DISTINTEGRATION OF GRANITE BY SALTS (AFTER KWAAD, 1970)

Treatment	Cycles	Weathering material produced from main block (g)
Dry	1	–
	2	–
	3	–
	4	–
Water	1	–
	2	–
	3	–
	4	–
$CaSO_4$	1	–
	2	–
	3	–
	4	–
$MgSO_4$	1	3.5
	2	3.6
	3	4.0
	4	2.7
Na_2CO_3	1	7.0
	2	5.3
	3	5.1
	4	7.7
Na_2SO_4	1	18.6
	2	2.6
	3	2.4
	4	4.6

Figure 2.8 Disintegration of clay bricks by sodium sulphate weathering at Mohenjo-Daro, Pakistan (After Goudie, 1977)

its weaker cement. The results of these authors are confirmed by those of Kwaad (1970) for granite (*Table 2.5*).

The environmental significance of sodium sulphate is stressed by Goudie (1977), who noted that this salt occurs in the ground water in Pakistan where it is a very effective agent in the disintegration of buildings constructed from mud bricks (*Figure 2.8*).

2.6 FREEZE–THAW

When water freezes it expands by about 10% of its volume. Thus water trapped in rock pores, joints and cracks will act to stress the rocks upon freezing. The arguments concerning the effects of this freezing relate to the role of water and hydration processes (White, 1976; Thorn, 1979). It is clear that the rock must first take up water into pores or into crystal structures and chemical compounds (hydration) and that the freezing water must be trapped within the rock body. Moreover, repeated freezing and thawing is necessary if progressive disintegration of the rock is to occur. Each time the water freezes the pores and cracks get larger, but following each thaw more water can enter the pores and cracks, filling them up and maintaining the pressures on the rock during the next freezing cycle. As a result it is usually the frequency of freeze–thaw cycles rather than the intensity of freezing that is most critical. Hydration may play a more important role than the freezing alone (White, 1976).

Rocks which have been broken up along joints and cracks are characteristically angular in outline, with sharp edges formed by mechanical fracturing and the absence of significant chemical action. Thus deposits of angular material (Briggs, 1977a, p. 106–107) usually attest to the existence, either currently or at some time in the past, of a climate with frequent frosts and associated cycles of freezing-and-thawing.

2.6.1 Method

Freeze–thaw conditions can be replicated by the use of a freezing compartment in a domestic refrigerator or by use of a deep freezer. The samples used will be similar to those employed in 2.2 and the cycle procedure similar except that the rock is frozen and thawed rather than wetted and dried. Placement for freezing can be carried out for 12 or 24 h as convenient and thawing should continue for 2–3 h, or until the rock is at room temperature. The sample may be placed in the refrigerator dry or it may be moistened. In practice, it is difficult to freeze a sample without some moisture or water vapour being present but a control experiment can be designed by drying one set of rocks in an oven and cooling them in a desiccator, while a replicate set are immersed in water for 12–24 h before freezing. This will illustrate the role of water in freeze–thaw processes.

The size of the sample may be a problem in that, in nature, freeze–thaw disintegration usually involves freezing down large joints and cracks of the sort not normally found in small hand-sized rock samples. Thus, it may only be possible to relate freeze–thaw disintegration to porosity in several small samples or to jointing or fracturing in one or two much larger samples.

2.6.2 Results

Results can be plotted either as mass loss (provided the sample is dried and cooled first) or as the number of fragments into which the rock breaks.

2.6.3 Case studies

Case studies are provided by the work of Wiman (1963) and Potts (1970). Up to 100 cycles were simulated and controlled temperature ranges of –8 °C to +8 °C were used.

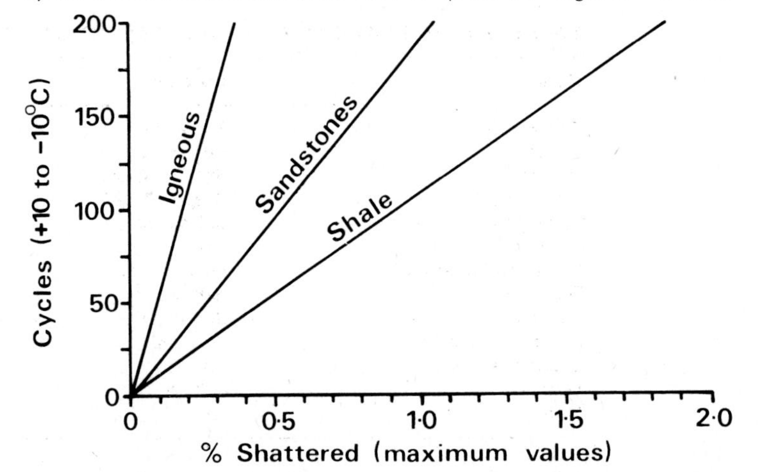

Figure 2.9 Susceptibility of rocks to freeze-thaw cycles of +10 to –10 °C in 24 h, maximum values recorded shown (After Wiman, 1963)

The samples which were immersed in water before freezing shattered more than those lightly moistened; lithology and grain size emerged as important factors and rates of shattering were related to the density of planes or points of weakness in the specimens, with shale shattering more than the other rocks studied (***Figure 2.9***). The number of oscillations across freezing point proved to be more important than the intensity of freezing. Nevertheless, the intensity of freezing needs to be controlled in experiments of this type. This may present a problem if freezing is being performed in a domestic refrigerator where precise temperature control is not always available. The temperatures used in experimentation should be recorded by placing a maximum–minimum thermometer in the freezing compartment used.

2.7 MECHANICAL STRENGTH OF ROCKS

In several instances in the above discussions it has been implied that the mechanical strength of the rock is an important factor in controlling rock disintegration. Relatively sophisticated devices, like the Schmidt hammer for concrete hardness, can be used for

TABLE 2.6 SCHMIDT HAMMER HARDNESS READINGS (FROM DAY AND GOUDIE, 1977)

Rock	Location	Mean R value
Precambrian quartzite	Botswana	67.10
Silcrete	Botswana	62.07
Red granite	Shap Fell, UK	61.10
White granite	Shap Fell, UK	59.38
Sarsen stones	S. England	54.90
Calcrete	Israel	51.54
Indurated limestone	Puerto Rico	53.00
Dolomite	Bahrain	50.00
Calcrete	Botswana	41.96
Dolomitic limestone	Sulawesi	37.50
Limestone	Jamaica	36.25
Aeolianite	Bahrain	14.45
Chalk	Israel	14.00

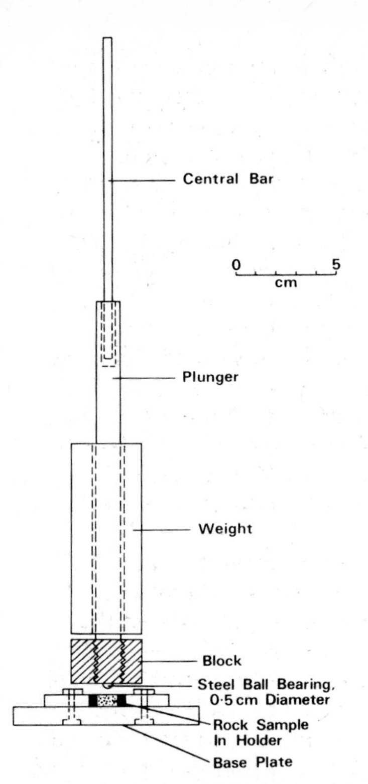

Figure 2.10 Device for dropping weights on rock for testing mechanical strength

testing the strength of rock material ***in situ.*** The concrete test hammer produces a controlled impact on the rock surface and the rebound value (*R*) depends upon the hardness of the surface. The higher the *R* value (on a scale of 10–100) the harder the rock and therefore, arguably, the greater its resistance to mechanical erosion. The test hammers are portable and are available from Engineering Laboratory Equipment (***see*** Appendix 1). Their use in a geomorphological context has been tested by Day and Goudie (1977) on a wide variety of rocks (***Table 2.6***).

For the small rock samples of the type already mentioned in this chapter, some of the mechanical engineer's devices can be adapted for use. The simplest of these is where a known mass is dropped from a known height onto samples of rock of standard size. This can be done very crudely, using weights from balances; for example, weights of increasing size can be dropped from a standard height until the rock fractures. A controllable device is illustrated in ***Figure 2.10.*** Here a weight is used with a central hole and this is slid onto a vertical rod. The rod is fixed at its lower end to a solid block under which the rock is placed. The mass can be varied but standard 1 kg or 0.5 kg weights are useful. The procedure is to increase the dropping height by 1 cm each time until the sample fractures, recording the height at which fracture occurs. This combines a cumulative stress with successively increased impact. An alternative is to drop repeatedly from a standard height and to record the number of drops necessary to induce fracture. A large number of samples from any one rock type should be used because the mechanical properties of rocks may vary considerably. Again, large samples cannot be tested but a comparative evaluation of rock types can be made.

2.7.1 Case study

Some results from the work of Smart (personal communication, University of Bristol) are shown in ***Table 2.7***. These show the variability which may be present in dolomitic lime-stone.

2.8 SOME GENERAL CONSIDERATIONS

The biggest limitation of all the experiments described in this section concerns the size of the sample used. Clearly, with many of the vessels or equipment used there is a limit on the size of sample it is possible to handle. However, the behaviour of large rock samples in nature may well be substantially different to the behaviour of small rock samples in experiments. Experiments can thus be used as a prediction of the behaviour of natural rocks, but not as a complete guide to their behaviour. The experiments may provide hypotheses which can be further tested by field observation. In the field, not only will larger rock masses be involved, but it will often be the case that several factors or processes will be operating simultaneously. Thus, the predictions from the laboratory should be tested by making observations on the weathering of rocks under natural conditions. Clearly, it will often be impossible to make observations on all the processes as they are happening because they often operate very slowly. However, the following could be noted in the field: What are the lithology and grain size of the rock? What are the likely agents of mechanical weathering (e.g. is it near a source of salt; do frosts occur?)? Are there any other factors involved (such as chemical weathering)? Is there any evidence of the occurrence of mechanical weathering? Is it likely to be active at the present day?

The investigator may make field observations and then enlarge upon them by laboratory experimentation of rock susceptibility or, alternatively, the laboratory experiments may be carried out and then field observations can be made to see if the experiments predict the field situation. For instance, in any one particular environment, it may be observed that screes are more extensive on one rock type than another. Samples may then be collected and put through freeze–thaw cycles in a fridge. If the results of the experiments coincide with field observations, then the theory that the scree material is produced by freeze–thaw action is not contradicted and the processes giving rise to the pattern observed can be more fully understood. If the results do not coincide then some other factors must be involved, for example differences in history of the sites or differences in chemical weathering. In any situation it is important to try to control the factors under experimentation and to realise the limitations of laboratory techniques when making inferences about the operation of natural processes.

TABLE 2.7 FRACTURE OF DOLOMITIC LIMESTONE

Sample (limestone)	Average weight drop height (cm) at which fracture occurred, 0.5 kg mass
1	8
2	6
3	9
4	10
5	5
6	10
7	3
8	6

CHAPTER 3 CHEMICAL AND BIOLOGICAL WEATHERING

3.1 INTRODUCTION

Chemical weathering involves the alteration of the mineral constituents of rocks by chemical reactions. Often these reactions are affected by biological processes, especially in soils, and it is difficult to separate biological and chemical processes in nature. However, even if the processes involved are biologically induced they can often be expressed and studied in terms of chemical reactions. As an example, minerals present in rocks can be weathered by acids and the weathering reaction can be studied in terms of the chemistry of the reaction between the acid and the mineral, but the acid itself may be derived from the physiological processes of organisms. Other weathering reactants include oxygen and water; in this chapter several weathering reactants are considered in turn, together with the effects of the reactants on different rocks.

The study of chemical and biological weathering involves four major considerations. Firstly, it is useful to evaluate the potential to facilitate weathering reactions; for example, the measurement of the acidity of water about to come into contact with a rock. Secondly, the susceptibility of individual rock types to any given set of weathering environments can be measured. Thirdly, the weathering residues left behind after chemical reactions have taken place can be studied and, finally, it is possible to study the removal of weathering products in flowing water after chemical reactions have taken place. This latter process constitutes chemical erosion and is discussed further in chapter 6; it is also an important control upon weathering reactions since further weathering will usually not occur until the products of previous weathering processes have been removed. Again, biological processes may be involved because weathering products may be removed from the site of weathering not only in inorganic solution, but also in combination with organic compounds or by uptake by plant roots. In the latter case, weathering products are being used as plant nutrients.

Some of the major processes of chemical weathering are illustrated in ***Figures 3.1–3.4.*** ***Figure 3.1*** illustrates the arrival of weathering reactants at a mineral surface. These reactants are adsorbed onto the surfaces where reaction takes place, followed by the

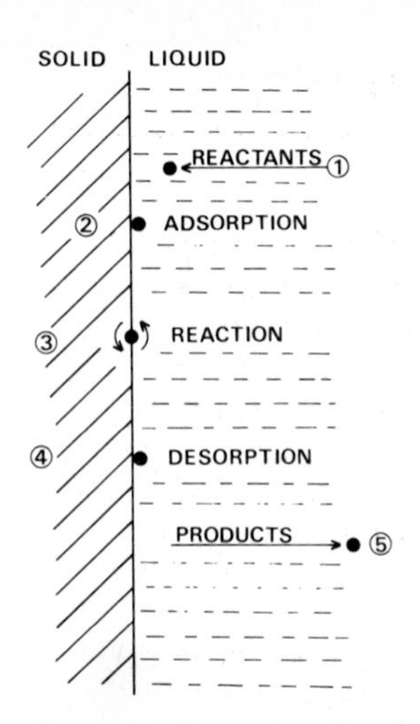

Figure 3.1 Illustration of surface processes in chemical weathering (After Mercado and Billings, 1975)

(*a*) H^+
  (1) $H_2O \rightarrow H^+ + OH^-$
  (2) ***ACIDS*** (a) $CO_2 + H_2O \rightarrow H_2CO_3$ (carbonic acid)
      $H_2CO_3 \rightarrow H^+ + HCO_3^-$
    (b) Organic acids
    (c) Mineral acids (e.g. H_2SO_4)
  (3) ***PLANT ROOTS*** H^+

(*b*) $O_2 \rightarrow$ OXIDATION

(*c*) ORGANIC ***CHELATES***
  (1) Leaf litter $\rightarrow$ Humus acids (Fulvic, Humic). Succinic, Lactic.
  (2) Rhizosphere micro-organisms (Root exudates) $\rightarrow$ Chelates.

Figure 3.2 Reactants in natural weathering processes

desorption of the products. Some examples of reactants are shown in ***Figure 3.2.*** The first group are the ***acids,*** derived from carbon dioxide, plant roots and other sources (***see*** section 3.2). The second reactant source is ***oxygen***, which leads to the oxidation of minerals. Thirdly, organic compounds from leaf decay and root activity produce reactants; here the reaction often involves ***chelation,*** that is the incorporation of a mineral ion into an organic structure (***see*** section 3.9).

Differential solubility is a common feature of chemical weathering, with the most soluble chemical elements being preferentially removed from the rock surface (***Figure 3.3***). Where water can penetrate into porous rocks and soil profiles, dissolution, transport and reprecipitation may occur (***Figure 3.4a***). Here weathering potential is used up as water flows through the porous medium and the products of reaction are moved through

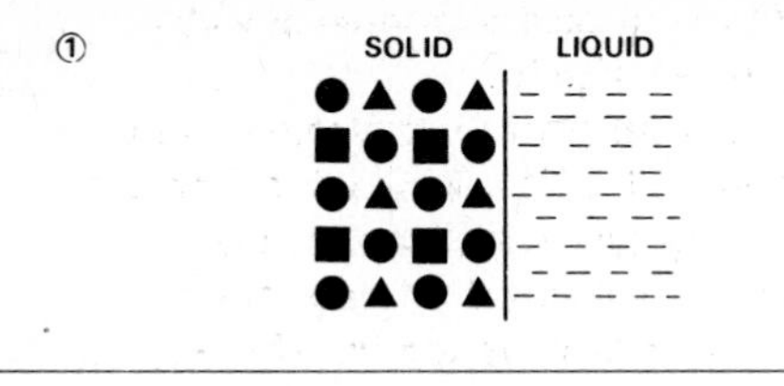

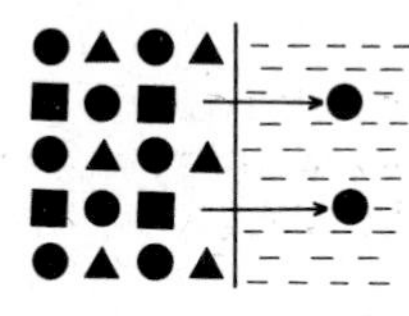

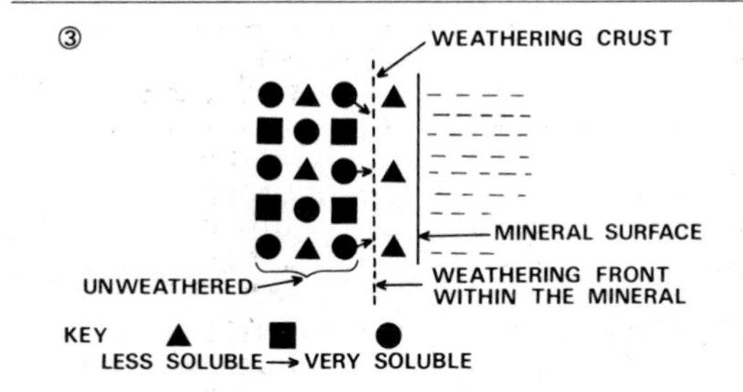

Figure 3.3 Differential solubility

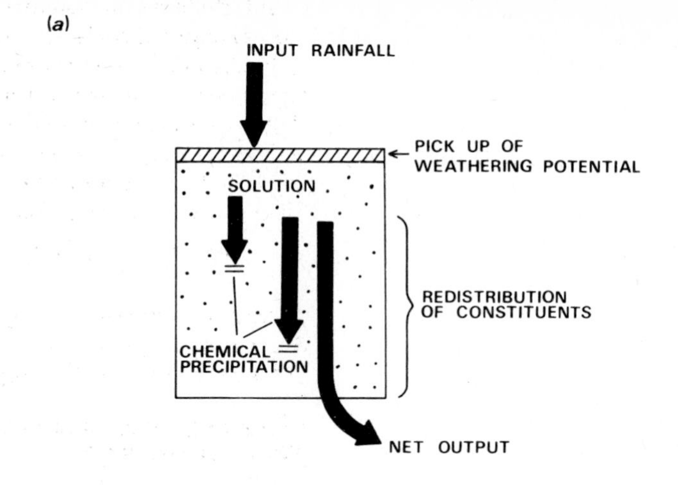

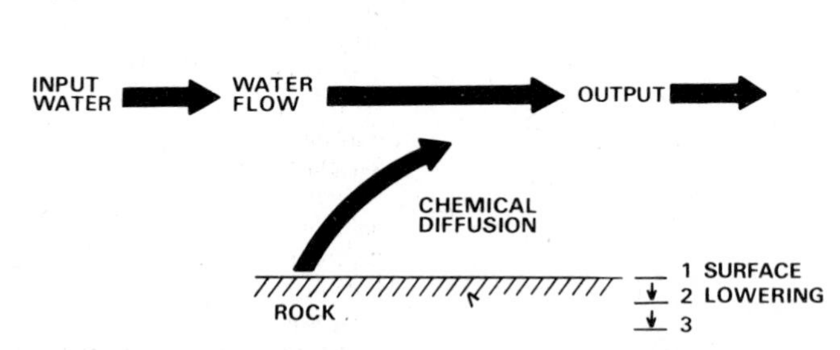

Figure 3.4(*a*) Porous system and (*b*) rock surface dynamics

or out of the system. On rock surfaces (***Figure 3.4b***) the surface is lowered as water moves across the surface, carrying reactants and products with it.

Predictions of whether or not a mineral is liable to weathering under earth surface conditions can be made using the approach of Curtis (1976). He calculated the energy changes involved in weathering reactions from the changes in the energies of the reactants and the products involved in the reaction. Chemical reactions involve energy transfer between the reactants and the products and the energy involved is usually referred to as Gibbs free energy, (G°), thus:

$$\Delta G^\circ_r = \Delta G^\circ_{fp} - \Delta G^\circ_{fr}$$

where ΔG° = change in energy, G°_r = standard free energy of the reaction, G°_{fp} = sum of the free energies of the products, and G°_{fr} = sum of the free energies of the reactants. If ΔG°_r is negative, then ΔG°_{fr} is higher than ΔG°_{fp}, and the

TABLE 3.1 ESTIMATES OF MINERAL STANDARD FREE ENERGY OF FORMATION VALUES (FROM CURTIS, 1976)

Mineral	Formula	ΔG°_f (Kcal mol^{-1})
Forsterite	Mg_2SiO_4	−491.9
Diopside	$CaMg(SiO_3)_2$	−752.8
Anorthite	$CaAl_2Si_2O_8$	−955.6
Albite	$NaAlSi_3O_8$	−884.0
Microcline	$KAlS_3O_8$	−892.8
Muscovite	$KAl_3Si_3O_{10}(OH)_2$	−1330.0
Fayalite	Fe_2SiO_4	−329.7
Wollastonite	$CaSiO_3$	−370.3
Spinel	$MgAl_2O_4$	−523.0
Jadeite	$NaAl(SiO_3)_2$	−677.2
Kyanite	Al_2SiO_5	−577.0
Pyrite	FeS_2	−38.3
Methane	CH_4	−12.1

TABLE 3.2 FORMATIONAL FREE ENERGY VALUES FOR POTENTIAL REACTANTS AND PRODUCTS STABLE IN THE WEATHERING ENVIRONMENT (FROM CURTIS, 1976)

Phase	Formula	ΔG°_f (Kcal mol^{-1})
Hydrogen ion	H^+ (aq)	zero
Sodium ion	Na^+ (aq)	−62.5
Potassium ion	K^+ (aq)	−67.7
Magnesium ion	Mg^{2+} (aq)	−108.9
Calcium ion	Ca^{2+} (aq)	−132.2
Hydrogen carbonate ion	HCO_3^- (aq)	−140.3
Sulphate ion	SO_4^{2-} (aq)	−177.3
Oxygen gas	O_2 (g)	zero
Water (liquid)	H_2O(l)	−56.7
Haematite	Fe_2O_3 (s)	−177.7
Quartz	SiO_2 (s)	−204.7
Gibbsite	$Al_2O_3 3H_2O$(s)	−550.3
Kaolinite	$Al_2Si_2O_5(OH)_4$ (s)	−904.0

more likely it is that a reaction will take place, and therefore the more unstable is the mineral.

Table 3.1 shows free energies of formation of various mineral phases and *Table 3.2* free energy values for potential reactants and products. If these values are ascribed to the reactions shown in *Table 3.3* the resulting values of ΔG°_r can be calculated as shown in *Table 3.4.* The values in *Table 3.4* are all negative, which suggests that there will be a tendency for the reactions shown to proceed. The suggestion is that the greater the negative value of ΔG°_r, the greater should be the tendency for the reaction to proceed (though it does not necessarily predict the rate at which it would proceed). Thus, for example, methane is highly unstable in the presence of oxygen, as is pyrite (reaction 12 in *Table 3.3*), predictable from a high negative value of ΔG°_f; in the feldspar group, microcline appears to be slightly more stable than anorthite during reaction with acids. These considerations can be used as a basis for an initial understanding of the relative order of mineral stabilities under earth surface conditions during specific reactions, and

TABLE 3.3 WEATHERING EQUATIONS WRITTEN WITH 'PRIMARY' MINERALS AS REACTANTS (FROM CURTIS, 1976)

1.	$Mg_2SiO_4(s) + 4H^+(aq) = 2Mg^{2+}(aq) + 2H_2O(l) + SiO_2(s)$
2.	$CaMg(SiO_3)_2(s) + 4H^+(aq) = Mg^{2+}(aq) + Ca^{2+}(aq) + 2H_2O(l) + 2SiO_2(s)$
3.	$CaAl_2Si_2O_8(s) + 2H^+(aq) + H_2O(l) = Al_2Si_2O_5(OH)_4(s) + Ca^{2+}(aq)$
4.	$2NaAlSi_3O_8 + 2H^+(aq) + H_2O(l) = Al_2Si_2O_5(OH)_4(s) + 4SiO_2(s) + 2Na^+(aq)$
5.	$2KAlS_3O_8(s) + 2H^+(aq) + H_2O(l) = Al_2Si_2O_5(OH)_4(s) + 4SiO_2(s) + 2K^+(aq)$
6.	$2KAl_3Si_3O_{10}(OH)_2(s) + 2H^+(aq) + 3H_2O(l) = 2K^+(aq) + 3Al_2Si_2O_5(OH)_4(s)$
7.	$Fe_2SiO_4(s) + \frac{1}{2}O_2(g) = Fe_2O_3(s) + SiO_2(s)$
8.	$CaSiO_3(s) + 2H^+(aq) = Ca^{2+}(aq) + H_2O(l) + SiO_2(s)$
9.	$MgAl_2O_4(s) + 2H^+(aq) + 2H_2O(l) = Al_2O_3.3H_2O(s) + Mg^{2+}(aq)$
10.	$2NaAlSi_2O_6(s) + 2H^+(aq) + H_2O(l) + Al_2Si_2O_5(OH)_4(s) + SiO_2(s) + 2Na^+(aq)$
11.	$2Al_2SiO_5(s) + 5H_2O(l) = Al_2Si_2O_5(OH)_4(s) + Al_2O_3.3H_2O(s)$
12.	$2FeS_2(s) + 4H_2O(l) + 7\frac{1}{2}O_2(g) = Fe_2O_3(s) + 4SO_4^{2-}(aq) + 8H^+(aq)$
13.	$CH_4(g) + 2O_2(g) = H_2O(l) + H^+(aq) + HCO_3^-(aq)$

aq — hydrated species in aqueous solution; l — liquid state; g — gas; s — solid

TABLE 3.4 GIBBS FREE ENERGY VALUES FOR WEATHERING REACTIONS (FROM CURTIS, 1976)

Mineral	ΔG°_f (Kcal mol^{-1})	ΔG°_f (Kcal g atom^{-1})
1. Forsterite	−44.0	−4.00
2. Diopside	−38.1	−2.72
3. Anorthite	−23.9	−1.32
4. Albite	−23.1	−0.75
5. Microcline	−15.9	−0.51
6. Muscovite	−17.3	−0.32
7. Fayalite	−52.7	−6.58
8. Wollastonite	−23.3	−3.32
9. Spinel	−22.8	−1.52
10. Jadeite	−27.3	−1.09
11. Kyanite	−16.8	−0.54
12. Pyrite	−583.5	−17.68
13. Methane	−184.9	−20.54

thus as a basis for hypothesis testing during experiments on mineral breakdown. It should be emphasised, however, that under natural conditions, other limiting factors may come into play, such as the supply of the reactants, limitations on the transport of products and the disposition of minerals in rocks and in relation to weathering surfaces. From a consideration of these general points, some specific examples of chemical weathering reactions can now be discussed.

3.2 WEATHERING POTENTIAL FROM ACIDS

Rocks and minerals tend to break down in the presence of acids – a process known as hydrolysis – and it follows that measurement of acidity will thus give an indication of the potential for weathering by acids. An acid is a compound which ***dissociates***, or splits into its constituent ions, in water to yield ***hydrogen ions***. These carry a positive charge and are written in the form H^+ (though they may actually be present in combination with the water molecule H_2O as the hydroxonium ion H_3O^+, but for convenience the acidity is expressed in terms of the excess H^+). A measurement of acidity is one which measures the amount of the H^+ ions in the water.

In pure water the H^+ ion is balanced by equal amounts of the hydroxyl ion, OH^-. This is taken to be a neutral state where the water is neither acid nor alkaline. Alkalinity is defined as the condition when a substance, termed an alkali, dissociates in water to yield a dominance of OH^- ions. Thus the relative concentrations of H^+ and OH^- ions in the solution determine its acidity or alkalinity. At the neutral state in pure water the hydrogen ion is present in a concentration of 0.000 000 1 g l^{-1}. Since this is a cumbersome number to deal with, it is simplified by taking the negative logarithm (to the base 10) of the number. This logarithmic expression is written $-\log_{10} [H^+]$ and is called the pH. In a neutral situation this yields a pH of 7. At high concentrations of H^+, acidity is higher but the logarithmic pH value is lower; for example 0.001 H^+ becomes pH 3. Thus a small pH value expresses high acidity and each change in pH unit expresses a tenfold change in hydrogen ion concentration.

Three examples of the production of acids in nature can be given: the dissociation of carbon dioxide in water; the production of sulphuric acid by bacteria; and the production of organic acids during the decomposition of organic matter.

Carbon dioxide is present in the atmosphere, making up 0.03% of its composition. When rain falls through the atmosphere it picks up carbon dioxide as the gas diffuses into the water droplets; similarly soil water may absorb carbon dioxide from soil air. In soil air the concentrations of carbon dioxide are relatively high due to its production by the respiration of soil organisms, including plant roots, and by the decomposition of organic matter (the carbon from the organic matter combining with the oxygen of the air). Carbon dioxide commonly constitutes 1–2% of the soil air. The gas forms carbonic acid in water which rapidly dissociates to yield hydrogen ions:

$$\underset{\text{gaseous carbon dioxide}}{CO_2} + \underset{\text{water}}{H_2O} \longrightarrow \underset{\text{carbonic acid}}{H_2CO_3}$$

$$\underset{\text{carbonic acid}}{H_2CO_3} \longrightarrow \underset{\text{hydrogen ion}}{H^+} + \underset{\text{hydrogen carbonate (bicarbonate) ion}}{HCO_3^-}$$

The hydrogen ion thus produced can then become involved in the hydrolysis of minerals.

The acid-producing bacterium ***Thiobacillus oxidans*** can oxidise sulphur to sulphuric acid:

$$\underset{\text{oxygen}}{O_2} + \underset{\text{sulphur}}{S} + \underset{\text{water}}{H_2O} \xrightarrow{\textit{Thiobacillus oxidans}} \underset{\text{sulphuric acid}}{H_2SO_4}$$

$$\underset{\text{sulphuric acid}}{H_2SO_4} \longrightarrow \underset{\text{hydrogen ions}}{2(H^+)} + \underset{\text{sulphate}}{SO_4^{2-}}$$

This reaction thus also yields hydrogen ions.

Organic acids are produced during the decay of organic matter and may include the more complex acids such as humic acid or fulvic acid or the simpler acids such as lactic, succinic or acetic acid. These may dissociate to yield hydrogen ions or become involved in organo-mineral reactions (discussed in section 3.8).

3.2.1 Method

Although the acids present in natural waters may be derived from a wide variety of sources, it is possible to assess the level of acidity by measuring the pH of the solution. Measurements can be made in a number of ways, including the use of coloured indicator solutions (which can be in the liquid form or impregnated onto absorbent papers) or electrically using a pH meter. In the case of the indicator solutions, the acidity present in the test solution alters the chemical configuration of the indicator and consequently its colour. The most widely used indicator is the BDH Universal Indicator which covers the pH range from 2 (acid) to 14 (alkaline), but other indicators are available which cover much narrower ranges (for BDH supplies, ***see*** Appendix 1). Electrically, pH may be measured by the potentiometric measurement of the electromotive force (e.m.f.) generated across a glass membrane by the concentration of hydrogen ions in an aqueous medium external to the membrane. The membrane forms the tip of an electrode which is plugged into a potentiometer calibrated in pH units.

Using any of these methods, one can readily assess the acidity of rain water, stream water, soil drainage water or sea water. In the case of indicator solutions, barium sulphate is used to flocculate (coagulate) clays in order to give a clear solution in which the colour is distinctive. In organic soils it may be necessary to add further indicator after settling, as organic matter may absorb the colour of the solution. When indicators are used the colour produced by the test solution is compared with a reference chart. With soils, pH may be measured by pressing a pH paper onto the moist soil surface. Alternatively a pH meter may be used and the soil may be moistened to give an adequate pH electrode–water contact. In very wet soils this is unnecessary, but with dry soils distilled water can be added either until the soil is in a sticky paste form (termed the 'sticky point' method) or a standard ratio of water to soil can be used, usually 1 to 2.5, e.g. 10 g soil to 25 ml water. Sources of material are given in Appendix 1 and methods are also described in Briggs (1977b). The choice of how to measure pH must take account of the aims of the study and the type of material being analysed.

3.2.2 Relative advantages of different measurement techniques

pH meters give more precise readings than Universal Indicator in solution or in papers, particularly in the central pH range. However, many more readings may have to be taken with a pH meter at any one site to determine average pH, for individual meter readings may simply reflect highly localised or temporary conditions; indicator methods, on the other hand, measure only the general, ambient pH of the sample. Because of variation in soil-moisture conditions, the use of a pH meter to measure soil acidity can yield results which may not be exactly reproducible from day to day. Moreover, although laboratory pH meters give the most instrumentally reliable results, pH can often change considerably upon transport and sample storage; drying and rewetting of the sample standardises any changes but removes the soil from its natural state.

Laboratory, mains pH meters should be used for experimentation and can be used for soils which have been dried and rewetted to a standard soil:water ratio. They may also be used for water samples provided that the samples are transported back to the laboratory in filled, sealed polythene bottles with tight-fitting caps, and analysed within a few hours. For field work portable pH meters are suitable for ***in situ*** measurements of water samples, and they may be useful for soil analyses (provided that the battery power is maintained at a stable level and the instrument is kept dry). Colorimetric tests are to be preferred, however, when less precise measurements of soil pH are required.

Some typical pH values measured by the author are given in ***Table 3.5.***

TABLE 3.5 pH VALUES OF NATURAL WATERS

Rain water (rural areas)*	pH: 6.5, 6.8, 7.1
Rain water (urban areas)*	pH: 6.5, 5.2, 6.1
Peat-bog water	pH: 4.3, 3.5, 4.7
Limestone streams†	pH: 7.2, 7.8, 8.2
Sea water ‡	pH: 8.2, 8.2, 8.1

*Rain water is slightly acidified by atmospheric carbon dioxide but may be made alkaline by mineral dust, or more acid by atmospheric pollution (especially when sulphur dioxide from fossil fuels combines with rain water to produce sulphuric acid).

†In streams draining limestone areas, chemical weathering has already taken place and thus the water is alkaline.

‡ Sea water is the repository of weathering products and is alkaline, normally at pH 8.2, but this may vary where evaporation occurs or where fresh water dilutes the sea water.

3.3 pH CHANGE DURING HYDROLYSIS

One of the main processes of chemical weathering is ***hydrolysis.*** During this process hydrogen ions are used up as they replace other ions in the mineral. As a result, the pH of the solution changes. These changes in acidity of the solution surrounding a weathering mineral can therefore be used to chart the loss of hydrogen ions during weathering. As an example, water acidified with carbon dioxide becomes less acid if it is in contact with limestone. The calcite of the limestone dissociates into its constituents which combine with the hydrogen ions as shown in ***Figure 3.5***. As the reaction proceeds, the acidity will drop until no further calcite can be dissolved because all the hydrogen ions in the solution are used up. Thus, the greater the initial supply of hydrogen ions – that is, the more acid the solution – the greater the amount of weathering that can take place.

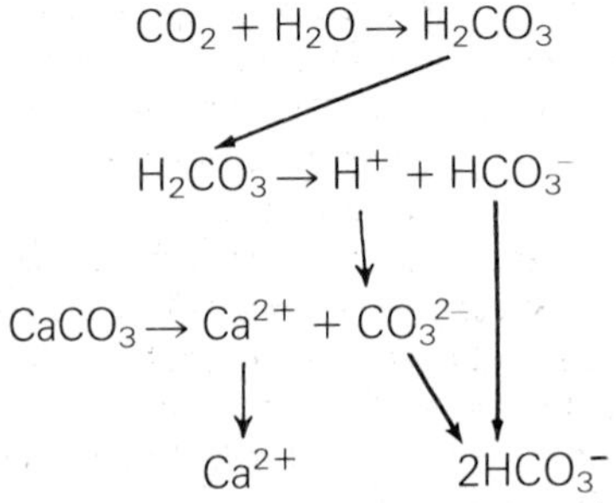

$$CO_2 + H_2O \rightarrow H_2CO_3$$

$$H_2CO_3 \rightarrow H^+ + HCO_3^-$$

$$CaCO_3 \rightarrow Ca^{2+} + CO_3^{2-}$$

$$Ca^{2+} \qquad 2HCO_3^-$$

Figure 3.5 Dissolution of calcium carbonate by hydrolysis and the production of calcium and hydrogen carbonate (bicarbonate) ions

3.3.1 Method

Several approaches to the analysis of hydrolysis can be adopted. Either several different rocks can be tested under standard conditions of acidity, or one or a few rocks can be tested in a wide variety of control or natural conditions. Alternatively, the experiments can be conducted on pure chemicals as well as, or instead of, rocks or minerals. The changes of pH during hydrolysis can be measured by any of the pH techniques available before, during and after the reaction.

3.3.2 Laboratory experiment

Aim

To monitor changes in pH during the hydrolysis of pure calcium carbonate in an acid solution.

Apparatus

a flask
tubing
distilled water
pure calcium carbonate (e.g. 10 g 'Analar' calcium carbonate)
pH meter or pH paper/indicator solution

Procedure

(1) Take 100 ml of deionised or distilled water. If the water has been standing in the laboratory it should already contain some carbon dioxide taken up from the air. Pure water which has equilibrated with atmospheric carbon dioxide will have a pH of 5.6. Alternatively, a tube may be inserted into the water and blown through so that exhaled carbon dioxide is bubbled through the liquid. A more effective way of evolving carbon dioxide is to place some calcium carbonate in a conical flask. Taking care not to spill the acid, a separate, small tube of dilute acid is then inserted in the flask. A stopper with some small-bore tubing is placed in the mouth of the flask and the other end of the tubing is placed below the water surface. The flask is then shaken to spill the acid onto the carbonate; carbon dioxide is evolved, and this bubbles through the water via the tubing.
(2) Measure the pH of the water; it should be acid because of the reaction described above (Step 1, ***Figure 3.5***).
(3) Stir in 10 g of powdered calcium carbonate powder and very quickly remeasure the pH.
(4) Continue to measure the pH until a stable reading is reached.

TABLE 3.6 RESULTS OF HYDROLYSIS EXPERIMENT WITH $CaCO_3$

pH of distilled water		5.6
After adding $CaCO_3$ and stirring		
	30 s	9.8
	2 min	10.0
	10 min	10.2
	20 min	10.2
	30 min	10.2

Note: If only indicator papers are available the pH changes may not show up very well as the colours are not strikingly different between pH 6.5 and 7.5, thus a pH meter should be used wherever possible. Indicator-solution colour standards may usefully be employed, however, by adding a few drops to a control beaker to which no calcium carbonate is added; this will facilitate colour comparison. Sample results from an experiment of this type are shown in ***Table 3.6.*** The pattern shown can be interpreted in terms of the reaction with H^+ ions described above (***Figure 3.5***).

3.3.3 Field method

Aim

To compare the susceptibility of rocks to weathering in natural waters of different weathering potential.

Apparatus

clean vessels (screw-top plastic or polythene bottles) for collecting water samples
funnels (polythene or plastic)
samples of ground-up rocks (crushed with a hammer)
pH meter or pH indicator

Procedure

(1) Collect a range of water samples from locally available sources (e.g. stream water and rain water). Rain water can be sampled by means of a plastic funnel in a bottle placed and secured in an open area.
(2) Measure the pH of the water samples (remembering that rain water in an industrial area may have been acidified by pollutants or in a rural area made alkaline by quarry dust, road dust or agricultural liming activity). The potential for weathering (acidity) can then be determined for each sample and ranked in order.
(3) Put 10 g of crushed rock into each sample, ensuring that the water samples are the same size.

Note: The procedure may be replicated as convenient (e.g. ten different water samples using one rock type or one large water sample subdivided amongst several rocks). The interpretation of pH changes may then be made in the light of the chemical environment of the source of the water and the nature of the rock. Sample results are shown in ***Table 3.7.***

TABLE 3.7 RESULTS OF SATURATION TEST

	pH	pH_2*	pH_3†
1. Rain water (non-industrial area)	6.8	7.6	6.8
2. Rain water (industrial area)	5.0	7.6	5.0
3. Stream water (polluted)	5.2	7.5	5.2
4. Stream water (upland from peat bog)	4.8	7.5	4.8
5. Stream water (lowland from chalky area)	7.5	7.4	7.5

*pH_2 – limestone added
†pH_3 – quartz added

3.3.4 Discussion

In urban areas, where acid rain may be present, some of the possible effects of this acidity on building stones may be evident. In the results shown in ***Table 3.7***, a clear reaction with limestone is seen, but no change occurs with quartz, which shows little susceptibility to hydrolysis. In sample 5 the stream was already alkaline and saturated with calcium carbonate and so the addition of limestone caused precipitation of calcium and a drop in pH (***see*** section 3.7 for a full explanation).

Caution is obviously required in the initial sampling of the waters in order that interpretations may take account of any possible contamination. Hypothesis testing can come from a knowledge of the minerals present in the rock capable of reacting with hydrogen ions. In general, it is those rocks which have carbonate (CO_3^{2-}) present in their chemical composition which will react in the way suggested.

3.3.5 Case study (Deju and Bhappu, 1965)

The authors dissolved several minerals in initially acid waters and used pH change due to hydrolysis as an indicator of the stability of minerals during weathering in acid environments. The results are shown in ***Figure 3.6***. As this illustrates, quartz is highly stable (not prone to hydrolysis) while olivine which contains magnesium (composition: $(Mg, Fe)_2SiO_4$) is unstable.

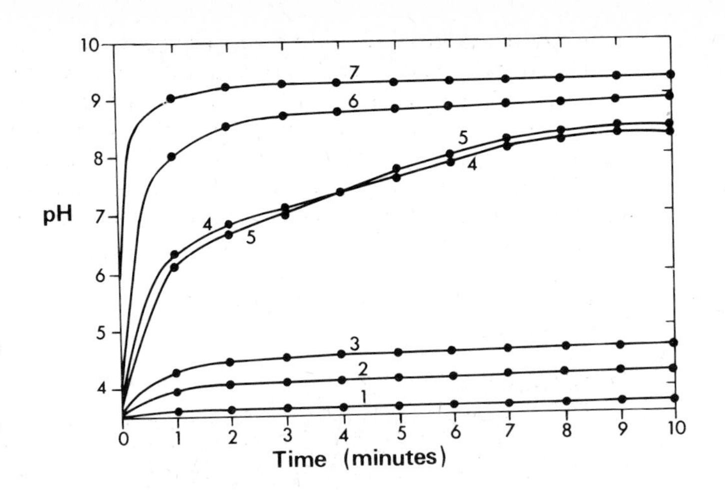

Figure 3.6 Hydrolysis of minerals, showing reduction of acidity during the process (After Deju and Bhappu, 1965 1 = quartz, 2 = albite, 3 = beryl, 4 = amphibole (actinolite), 5 = hornblende, 6 = pyroxene (enstatie – augite), 7 = olivine (forsterite)

3.4 RELEASE OF CHEMICAL ELEMENTS

Chemical weathering releases chemical elements into solution, where their presence can often be detected by various testing techniques. Both spot tests for presence/absence and also quantitative assessments can be used. Under controlled conditions, measurements of the amounts of specified elements in solution at equilibrium can be used to indicate the solubility of the elements in question.

There are considerable differences in the levels of sophistication possible in chemical analyses. At the simplest level, colour changes or the formation of a precipitate can be used; at a slightly more sophisticated (although not necessarily expensive) level, titrations can be carried out. For more detailed experiments, quantitative analyses can be performed using flame photometers or spectrophotometers. The treatment in this book is necessarily selective. Reference should be made to BDH (1973) for the simpler methods and to Allen (1974) for further methods.

Spot tests can be used to ascertain whether or not a chemical element has been released into solution during a weathering reaction. In some cases the intensity of colour development may be used as a guide to the amounts of element present. Spot test kits are available from Camlab. Ltd (Appendix 1) for some chemical elements.

The tests can be used on ground rocks, rock fragments, soils or water samples. Distilled water or acid can be added to the rock or soil sample. The solution should then be allowed to settle, or it should be filtered, to obtain a clear solution. The water or acid should also be tested to check that the test substance does not exist in the solution before it is added to the rock or soil.

3.4.1 Calcium spot test

Calcium can be derived by the hydrolytic weathering of limestones and other calcium-bearing rocks.

Procedure

(1) Select a range of rocks of known lithology and, using either ground or crushed rocks or rock fragments, place them into about 10 ml of distilled water.

(2) Add a few drops of hydrochloric acid (the concentration is not critical, but 10% v/v is adequate).

(3) Wait a few minutes and then add five drops of saturated ammonium oxalate solution.

N.B. The ***oxalate is poisonous*** if taken internally.

If calcium is present in solution then a white precipitate of calcium oxalate will be visible.

3.4.2 Quantitative analysis for calcium and magnesium in solution

This technique can be used for the analysis of stream waters, soil extracts, or distilled waters or acids in which rocks have been dissolved. In the last case, known amounts of rock can be added to known solution volumes. Analysis kits are available from some suppliers (see Appendix 1) and are also described by Douglas (1968). A technique for titration is outlined below.

Background

Titration with EDTA is a standard technique for the quantitative analysis of Ca^{2+} and Mg^{2+} ions in solution. EDTA (ethylenediamine tetraacetic acid, in the disodium salt form) forms stable complexes with a large number of metallic cations (calcium, magnesium, iron, cadmium, zinc, mercury, copper, cobalt, silver, nickel, aluminium, platinum, titanium, beryllium, manganese and lead). In karst waters it is normal to assume that calcium and magnesium are the dominant cations present, but in heavily contaminated waters and in the analysis of soils and rocks it is necessary to use masking substances in order to inactivate the interfering elements. Triethanolamine masks iron, manganese and copper; sodium (D+) tartrate masks aluminium; hydroxylammonium hydrochloride masks iron, manganese and copper; and potassium cyanide masks copper, cobalt, nickel and zinc but is not normally used because of its poisonous nature. Sodium tartrate is harmful by skin adsorption. A 50% triethanolamine solution is a recommended masking agent.

The standard procedure is to titrate two ***aliquots*** (subsamples) from the same sample, one for calcium and one for calcium + magnesium. This is achieved by adding a strong alkali to one aliquot which has the effect of precipitating the magnesium so that it does not take part in the titration. The titration for calcium + magnesium is often referred to as the total hardness and the figure for magnesium is gained by subtracting that for the calcium from the total.

The calcium titration proceeds by the use of potassium hydroxide, to alkalise the solution and precipitate the magnesium, and the addition of an indicator which forms a complex with the free calcium. When the EDTA is added to the solution, the calcium is complexed by the EDTA (which forms a stronger complex than the indicator–calcium complex) and, upon losing calcium, the indicator changes colour. When the colour change is complete, the end point has been reached and the amount of EDTA used is noted. The amount of calcium present in the solution is computed from a knowledge of the ratio in which calcium and EDTA combine.

The calcium + magnesium titration is undertaken on the same principles but a mildly alkaline solution is used instead of a strongly alkaline one.

Apparatus and chemicals needed

100 ml measuring cylinder or pipette (or use a 50 ml one twice)
250 ml conical flask

retort stand, bosshead and clamp
white tile or white piece of paper
100 ml or 50 ml burette
0.025M EDTA (9.306 g ℓ^{-1})
Ammonium purpurate indicator for calcium titration (0.02 g in 10 ml distilled water)
Potassium hydroxide buffer for calcium titration (80 g ℓ^{-1}) (*CAUSTIC*)
Erio-T indicator for calcium + magnesium titration (0.02 g in 10 ml alcohol or 50% triethanolamine)
Buffer for calcium + magnesium titration (70 g ammonium chloride in 570 ml of concentrated ammonia and made up to 1 ℓ in a fume cupboard)
2 small beakers
pH paper
small funnel

Procedure

CAUTION: POTASSIUM HYDROXIDE IS CAUSTIC TO THE SKIN AND AMMONIA CAUSES IRRITATION TO EYES AND LUNGS.

(1) Set up retort stand with clamp and bosshead; clamp in burette so that tip is just above level of 250 ml flask, placing a white tile or clean paper under the flask to facilitate estimation of colour change.
(2) Remove flask from under burette.
(3) Fill burette with EDTA using the small funnel, taking care not to overfill funnel.
(4) Check that tap flows freely.
(5) Rinse out the measuring cylinder, or pipettes, flask and beakers with distilled water.

Calcium

Prepare two colour standards. Take approximately 25 ml of sample, add potassium hydroxide buffer to pH 14 (use pH papers to check) and ammonium purpurate indicator (a few drops to give a good colour). Repeat this in the other beaker.
To one beaker add EDTA until a strong colour change occurs. The beaker without

EDTA should be a reddish purple colour. The one with an excess of EDTA should be pure mauve without any trace of red.
Place these beakers one either side of the flask and use these as colour reference points to ascertain the end point during the titration.
Titrate the sample. Take 100 ml of the water sample, measured in the measuring cylinder or pipette. Pour or pipette into the conical flask.
Add potassium hydroxide to pH 14.
Add a few drops of indicator to colour.
Perform a rough titration first and then an accurate one. Note the burette reading and add EDTA until the colour changes. Note the burette reading and compute the EDTA used.
Prepare a fresh 100 ml of sample and add buffer and indicator as appropriate. Note the burette reading (R_1).
Run in most of the EDTA to be used but approach the end point carefully adding the last amount of EDTA accurately, drop by drop (if you are unsure whether the end point has been reached, read the burette before adding the next drop). R_2 is the final burette reading at the end point.
Compute:

$$R_2 - R_1 = V_1 \quad (V_1 = \text{ml EDTA used in titrating calcium})$$

$$V_1 \times 25 = CaCO_3 \text{ content of water sample in mg } \ell^{-1}$$

Magnesium
Prepare two colour standards as above, but using ammonia buffer to pH 10 and a few drops of Erio-T indicator. Before addition of EDTA the solution should be purple and after addition it should be a clear blue.
Perform a rough titration as before and then an accurate one, approaching the end point dropwise.
Read burette before titration (R_3) and after (R_4), then calculate:

$R_4 - R_3 = V_2$ (ml EDTA used on calcium + magnesium)

$V_2 - V_1 = V_3$ (ml EDTA used on magnesium alone)

$V_3 \times 21.1 = MgCO_3$ content of water sample in mg ℓ^{-1}

Write up environmental interpretation of results.
Notes Burette reading can be facilitated by placing a spliced piece of paper on the burette column.
With very low concentrations of magnesium it is best to adopt the following procedure: Take R_3, add most of the EDTA amount used in calcium titration, add buffer and indicator, then complete titration dropwise.

3.4.3 Phosphate test

This test can be used to demonstrate the presence of phosphate in some rocks, for example, limestones.
(a) ***Spot test.*** (1) Acidify a rock and water mixture with sulphuric or hydrochloric acid.
(2) Add a few crystals of ammonium molybdate and some of tin(II) chloride (stannous chloride). A blue colour indicates the presence of phosphate. (Take care to check solutions used without rock, as phosphate may be derived from plastics or detergent.)
(b) ***Quantitative test. Reagent*** (i) – Dissolve 6.5 g of ammonium molybdate in a solution of 110 ml concentrated sulphuric ***acid added to 500 ml distilled water*** (*N.B.* water added to concentrated acid may result in an explosion). Add 3.5 ml to sample.
Reagent (ii) – Dissolve 40 g tin(II)chloride to a solution of 50 ml concentrated hydrochloric ***acid added to 50 ml distilled water*** (this is the stock solution). Add 1 ml of the stock solution to a solution of 5 ml concentrated hydrochloric acid in 100 ml distilled water (make this up freshly from stock for each set of determinations as it is unstable). Add 0.2 ml to test solution.
Standards – Dissolve 13.609 g of potassium dihydrogen orthophosphate in 1 ℓ of distilled water. This gives 100 mg ℓ^{-1} phosphate. Dilute to give 50, 25, 10, mg ℓ^{-1} etc., and/or use 1.3609 g ℓ^{-1} to give 10 mg ℓ^{-1} and dilute to get 5, 2, 1 mg ℓ^{-1}. To obtain 50 mg

ℓ^{-1}, for example, take 50 ml of the initial standard and dilute to 100 ml with distilled water. Add the quantities of the reagents specified to the standards of known concentrations and compare the colours developed with those of the sample solutions. Reagents (i) and (ii) can be substituted by Tschopps Reagents 1 and 2 from BDH. Colour development can be measured using a BDH colour comparator, a colorimeter or a spectrophotometer.

3.4.4 Other chemical tests

Many other chemical tests can be carried out, according to the range of apparatus available. Simple kits are available from BDH and Camlab (***see*** Appendix 1).

3.5 OXIDATION AND REDUCTION

Oxidation is an important chemical weathering process. It takes place when minerals which were formed in an anoxic (i.e. oxygen-deficient) environment are exposed to the atmosphere. Under these conditions, minerals such as pyrite (FeS_2) react with oxygen in the air and undergo chemical alteration. The potential for oxidation is represented on an oxidation--reduction scale (***Figure 3.7***), known as the Eh scale. ***Oxidation*** involves the loss of an electron. Since electrons are negative then losses can be expressed as an increase in the valency of the element involved in the reaction. The reverse process is known as ***reduction***, which is the gain of negative electrons, and thus involves a reduction in the net charge of the ion:

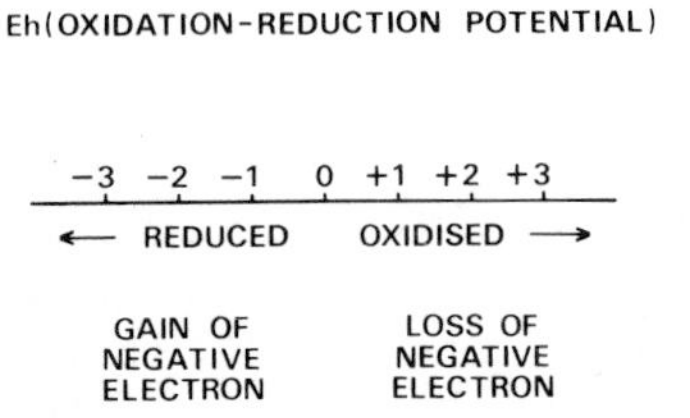

$$\underset{\text{(trivalent iron)}}{Fe^{3+}} + \underset{\text{(electron)}}{e^-} \underset{\text{oxidation}}{\overset{\text{reduction}}{\rightleftharpoons}} \underset{\text{(divalent iron)}}{Fe^{2+}}$$

Eh(OXIDATION-REDUCTION POTENTIAL)

−3 −2 −1 0 +1 +2 +3

← REDUCED OXIDISED →

GAIN OF NEGATIVE ELECTRON

LOSS OF NEGATIVE ELECTRON

Figure 3.7 Oxidation—reduction scale

Oxidation is therefore represented by increasing positive values on the Eh scale and reduction by increasing negative values (***Figure 3.7***).

Iron is the mineral element most commonly involved in oxidation but other elements may be involved, especially manganese. Iron(II) (ferrous iron, Fe^{2+}) is the reduced from of iron and is divalent. It is relatively unstable and readily oxidises to iron(III) (ferric iron, Fe^{3+} when exposed to the atmosphere; iron(III) is trivalent.

Oxidation and reduction are not affected only by the availability of oxygen: pH can also have an effect. Iron may be oxidised at high pH values even if the Eh scale is negative (reducing). The relation of the Eh scale to the pH scale is shown in ***Figure 3.8.***

IRON MOBILISED AT LOW pH (HIGH ACIDITY)
AND WHEN REDUCED (LOW Eh)

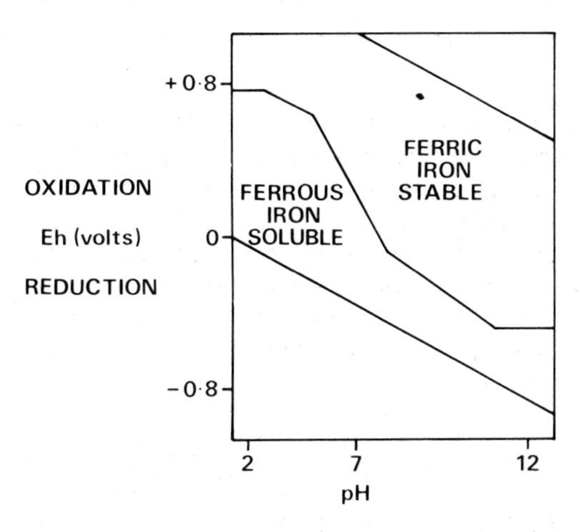

FERROUS IRON ; SOLUBLE (IRON (II), Fe^{2+})
LOW pH, Eh +0·8 ⟶ 0
HIGH pH, Eh 0 ⟶ −0·8

FERRIC IRON ; STABLE (IRON (III), Fe^{3+})
LOW pH, Eh > +0·8
HIGH pH, Eh +0·7 ⟶ −0·7

Figure 3.8 Eh–pH relations

Measurements of Eh are undertaken with an Eh electrode. This is commonly a platinum electrode which is sensitive to differences in Eh in the environment. The oxidation–reduction reaction at the surface of the electrode leads to the formation of an electrical potential which can be measured potentiometrically using an Eh meter or by attachment of the electrode to a pH meter with a millivolt scale. If such relatively expensive apparatus is not available then some simple field tests can be used to indicate the presence of ferrous and ferric iron. First it should be noted that iron(III) is usually red or brown in colour (i.e. rusty) and that iron(II) in soil may give it a pale bluish, greyish or greenish colour.

3.5.1 Method

Procedure for soils

(1) Place small amounts of fresh soil (e.g. a penknife-tipful or small spatula-tipful) into two separate small beakers or onto a white tile.
(2) Moisten the samples with dilute (1M) hydrochloric acid.
(3) To one sample add potassium thiocyanite solution (10% weight/volume e.g. 10 g per 100 ml). Any iron(III) present will turn red.
(4) To the other soil sample add potassium hexacyanoferrate(III) (potassium ferricyanide) solution (0.5% weight/volume, e.g. 0.5 g per 100 ml); Iron(II) will turn blue.

Note: It is useful to sprinkle the stained soil onto a white powder such as kaolin or barium sulphate to show up the colour.

The method can also be used on rock chippings, weathered and unweathered rock surfaces and on cut rock faces. For carbonate rocks the indicators may be mixed with an acid as follows.

Procedure for limestones

(1) Prepare 1.75% HCl (add 50 ml concentrated hydrochloric acid slowly to 750 ml distilled water in a fume cupboard). DO NOT ADD THE WATER TO THE ACID.
(2) Take 250 ml of this and add 4.2 g potassium hexacyanoferrate(III) in a fume cupboard.
(3) Dip part of the rock in this solution for 1 min. A blue colour indicates the presence of iron(II) and a red colour of iron(III). The latter is more stable and resistant to

weathering under conditions at the earth surface. (Potassium hexacyanoferrate(III) in a weak hydrochloric acid solution is relatively stable and does not lead to cyanide gas evolution, but nevertheless the solution should be handled with care and ***only*** treated in the way given.)

3.6 WEATHERING-POTENTIAL TABLETS

During weathering, rocks undergo changes in mass as their less stable constituents are removed. One way of monitoring weathering, therefore, is to measure the changes in mass. Rock specimens can be weighed and placed in various environments and any mass loss can be measured upon the retrieval of the samples (***Figure 3.9***). The samples have to be weighed very accurately, preferably to four or six decimal places of a gram. Mass loss of soluble rock material can be used as an indicator of weathering potential in an environment; for example samples of limestone rock can be placed at the base of two soil profiles, one an acid soil and one an alkaline. Mass loss after one year's emplacement would be expected to be greater under the acid soil. At least 30 samples should be placed under one environmental condition if meaningful, statistically valid comparisons are to be made, and they should preferably be placed at randomly chosen sites.

Both naturally occurring fragments and specially cut tablets can be used. The advantage of using cut tablets is that standard conditions are maintained; all tablets have the same surface area, the same size and the same shape. It may be convenient to produce tablets by coring a piece of rock and then slicing this on a rock cutter. Alternatively cubes may be cut with a rock-cutting machine. It can be argued that an unnatural surface is being used here, and that a faster rate of mass loss may be engendered by the use of a cut surface. However, since the main purpose of the exercise is to compare sites using a standard technique, the nature of the surfaces is not crucial so long as they are all identical. It is also possible, however, to collect stones of roughly similar size, to remove any dirt and to use these.

Weighing should be performed on a balance accurate to at least 0.000 1 g. Measurable mass losses can be increased by leaving the tablets in for longer or by using highly soluble material, such as limestone. Clearly, the most detectable losses will then be incurred under the most acid conditions.

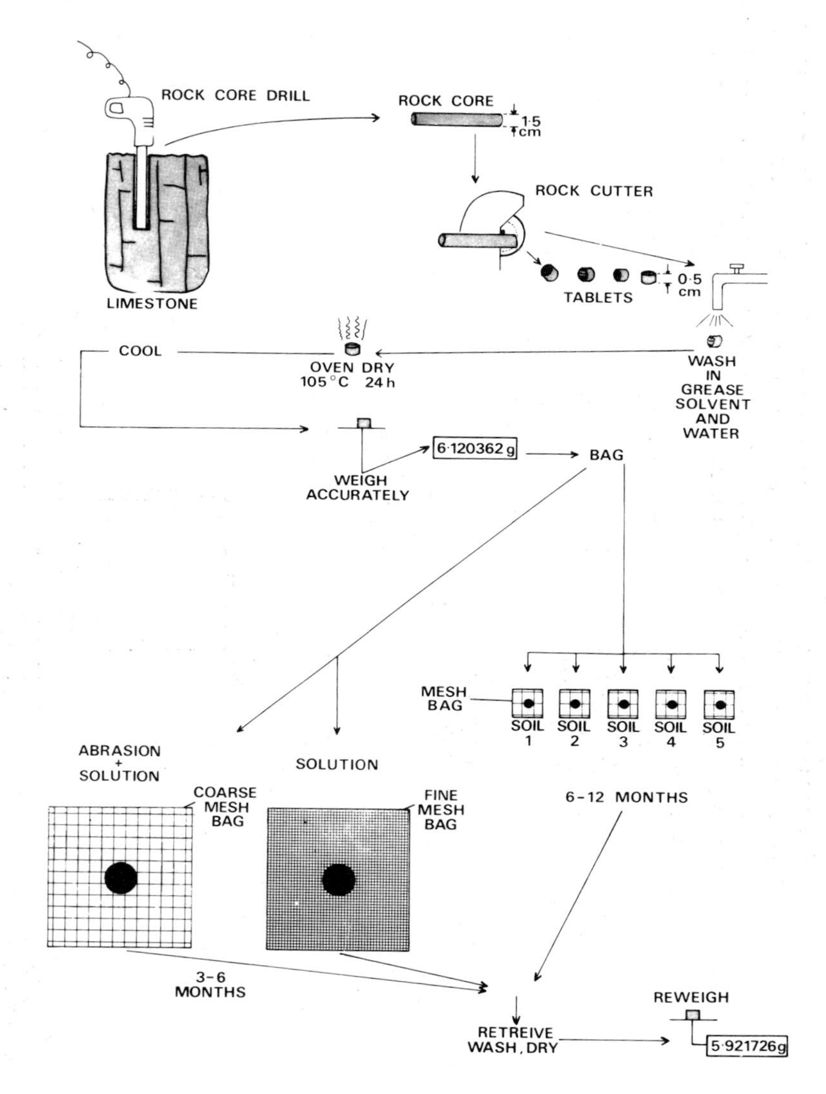

Figure 3.9 Rock tablet mass-loss measurement procedure

Care must be taken to ensure that the tablets may be identified upon retrieval. They can be laid out in numbered order (say from east to west, or down a soil profile) and coloured pegs may be used to mark the site. In vigorous environments the stones may be bagged using coarse mesh nylon bags, tied with nylon line (***Figure 3.10***).

When tablets are placed under soils a problem may exist if some soil adheres to the tablet upon retrieval. This can be obviated by the use of nylon mesh bags but this, in

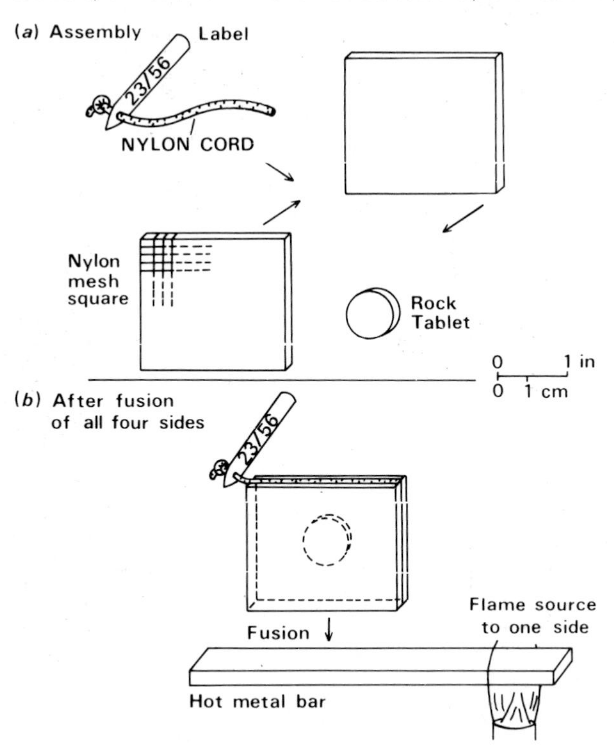

Figure 3.10 Rock tablet bagging for vigorous environments (After Trudgill, 1975)

turn, may result in a poor contact between the tablet and the soil solution unless care is taken to ensure a good contact during insertion. If some soil does adhere to the tablet, it may normally be removed by light brushing. The best rock tablets to use are those with sound surfaces where soil particles may not become lodged within any holes on the surface. In most cases, emplacement periods of at least one year will be required to gain appreciable mass losses which will be greater than those incurred by any treatment method. Details of the method are given in Trudgill (1975).

On a limestone bedrock, calcareous soils tend to protect the rock from weathering since most of the weathering potential is dissipated in the calcareous material of the soil. Acid soils provide considerable weathering potential at the soil–bedrock interface (***Figure 3.11***). These tendencies can be tested by placing mass-loss tablets in contrasting soil

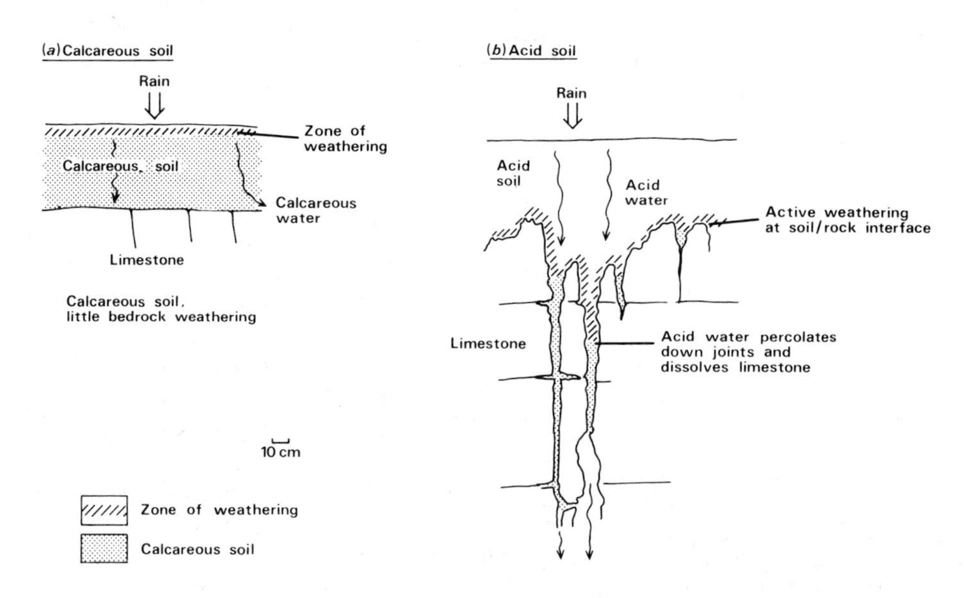

Figure 3.11 Weathering potential for limestone under calcareous and acid soils (After Curtis, Courtney and Trudgill, 1976)

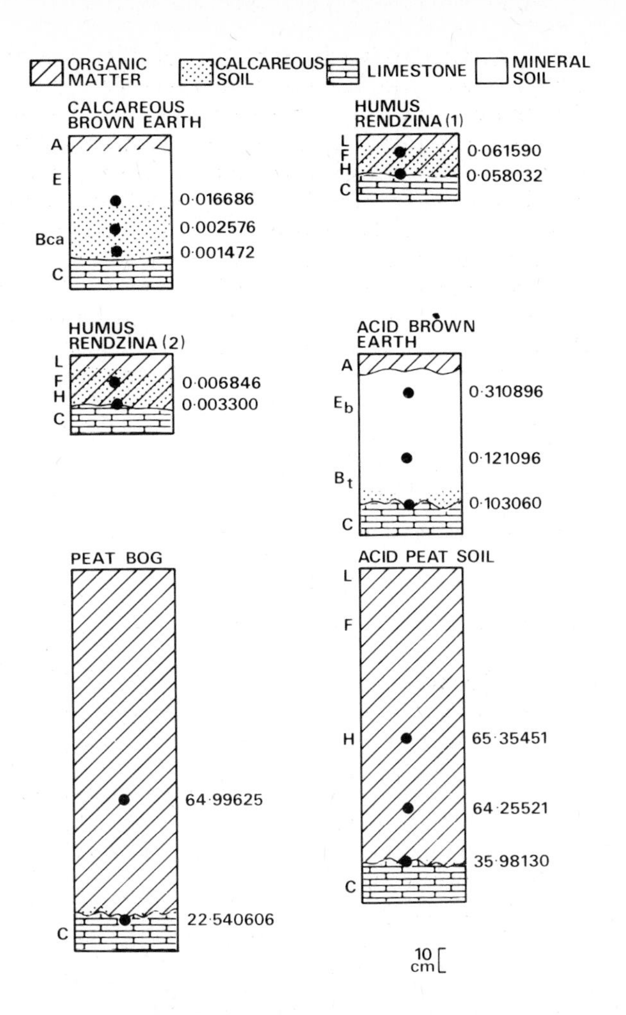

Figure 3.12 Mass-loss tablet results for limestones under a variety of soil types (mg a^{-1}) (After Trudgill, 1976a, c)

situations (***Figure 3.12***). The highest rates of bedrock erosion occur under the most acid soils, but in the calcareous soils, although the bedrock is protected, solutional erosion and surface lowering still occur, except that they take place within the soil. Care must be taken to place the tablets at the soil–bedrock interface, or on top of the zone of carbonate-rich soil, rather than in acid 'A' horizons of soil. If placed in the latter situation the tablet results will be related simply to soil acidity and will not be relevant to naturally occurring surface-lowering processes.

3.7 THE AGGRESSIVENESS OF WATER

The term aggressiveness is used to indicate the potential which water possesses for dissolving calcium carbonate. Measurements of aggressiveness are undertaken when water is entering a carbonate system of interest so that the likely effect of that water on dissolution processes can be assessed. Changes in aggressiveness during the passage of water through a carbonate system, such as a limestone soil profile or a cave system, can be used to plot the development of the solution processes.

There are two chief ways of measurement: (1) by pH change and (2) by double titration. Both methods rely on the addition of powdered calcium carbonate to a water sample.

3.7.1 pH change

Procedure

(1) Measure the pH of a water sample immediately on sampling.
(2) Add about 1–2 g calcium carbonate powder to the sample retained in the pH electrode holder or to the sample, if held in a bottle.
(3) Note any pH change.

A decrease in pH upon the addition of calcium carbonate powder means that the powder has seeded the solution with crystallisation foci and that precipitation has taken place around these from a supersaturated solution; an increase in pH means that calcium carbonate has gone into solution from the added powder, and thus that the original sample had the ability to take up further calcium carbonate (i.e., that the solution was aggressive). No change in pH indicates that the solution was already at saturation. This procedure and the reactions of the sample can be understood with reference to the carbonate equilibria outlined on p. 59. Simply expressed, during the dissolution of the powdered calcium carbonate, hydrogen ions are used up, this makes the pH shift from acidity to alkalinity. The degree of pH change is therefore a measure of the free hydrogen ions in solution capable of hydrolysis of calcium carbonate.

The use of calcium carbonate powder provides a standard method whereby comparisons can be made between one water body and the next. It is often the case, however, that actual limestones do not behave in the same way as pure calcium carbonate because limestones contain trace elements and other impurities. Thus, it is important to repeat the measurement using a sample of crushed limestone from the area in question. Care must be taken to make the limestone powder roughly the same grain size as the calcium carbonate powder used initially, otherwise the reaction will not take place so rapidly. To crush the rocks in the field, the rock can be placed in a strong polythene bag and broken with a geological hammer until some rock dust is formed. It is preferable to grind a sample of rock in the laboratory and to pass the material through sieves, retaining the fraction between 60 and 200 μm and to use this standardised rock powder to add to the field water sample.

3.7.2 Double titration

Again, powdered calcium carbonate is added to a water sample but in this case it is the changes in calcium content which are monitored rather than changes in pH.

Procedure

(1) Take two water samples in separate bottles simultaneously from one field site.
(2) Analyse one of these – the 'natural' sample – for calcium in solution (***see*** section 3.4.3).
(3) Add 1–2 g calcium carbonate powder to the second sample immediately after sampling. Retain for analysis in a tightly stoppered bottle for at least 8 h but not more than 48 h; 12–24 h is the optimum time span, during which saturation is likely to be achieved but degassing of carbon dioxide is liable to be minimised. Once more, powdered rock samples may be used to saturate a further water sample. The sample bottle should be filled to the top and must be kept tightly stoppered while the

reaction is taking place. If the calcium content of the natural sample is less than that of the artificially saturated one then calcium has gone into solution from the powder and the original sample was aggressive. Degrees of aggressiveness can be measured in a range of samples by comparing amounts of calcium going into solution. As with pH

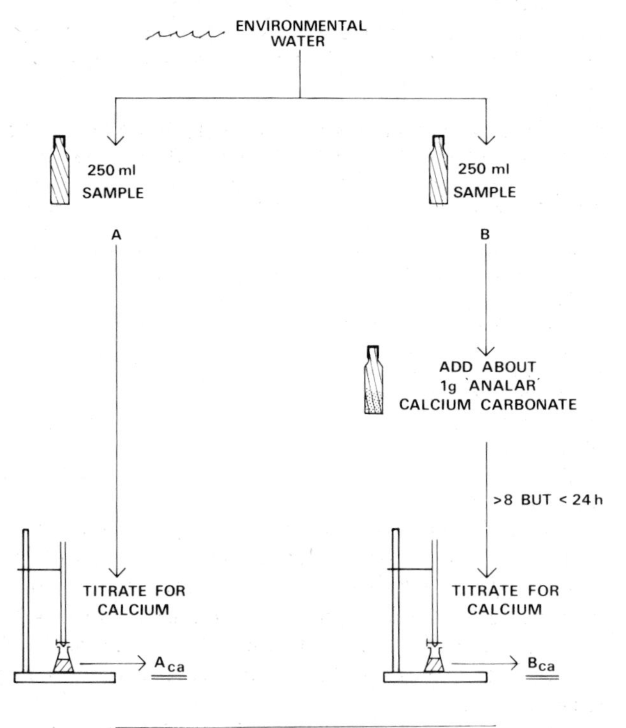

Figure 3.13 Scheme for aggressiveness titrations for carbonate waters

measurements, no change indicates that saturation already exists and a decrease indicates that precipitation has occurred from a supersaturated solution. The scheme of analysis is shown in ***Figure 3.13.***

3.7.3 Case study

Hypothesis

That in a limestone drainage system, aggressiveness is derived from drainage on non-carbonate rocks and is used up during the hydrolysis of limestone. The data are shown in *Table 3.8.*

TABLE 3.8 CHANGES IN CARBONATE CHEMISTRY OF WATERS ENTERING AND LEAVING A LIMESTONE SYSTEM, PEAK CAVERN, DERBYSHIRE, UK

Sampling station	pH	pH SAT*	$CaCO_3$ (mg ℓ^{-1})	$CaCO_3$ SAT (mg ℓ^{-1})	Conclusion
1. Stream running off shales, Rushup Edge	6.5	7.2 (+0.7)	40	70 (+30)	Aggressive
2. Stream sinking into limestone at Perryfoot Swaller	7.0	7.2 (+0.2)	60	70 (+10)	Aggressive
3. Stream emerging from limestone at Peak Cavern	8.2	8.1 (−0.1)	110	90 (−20)	Saturated

**see* p. 80

Interpretation

In this locality, aggressive water is running off shales and picking up calcium before it reaches the swallet site, where the stream then sinks into the limestone. At the resurgence, where the water re-emerges from the limestone, the water is saturated with respect to calcium carbonate. Saturation has occurred because of hydrolysis of limestone; this will include dissolution of cave walls but also there will have been large additions of saturated calcium carbonate solutions derived from percolation through the bedrock above the cave.

A further illustration is given in ***Figure 3.14,*** for samples taken on two separate days downstream of a site where peat-bog waters drain onto Carboniferous Limestone. While

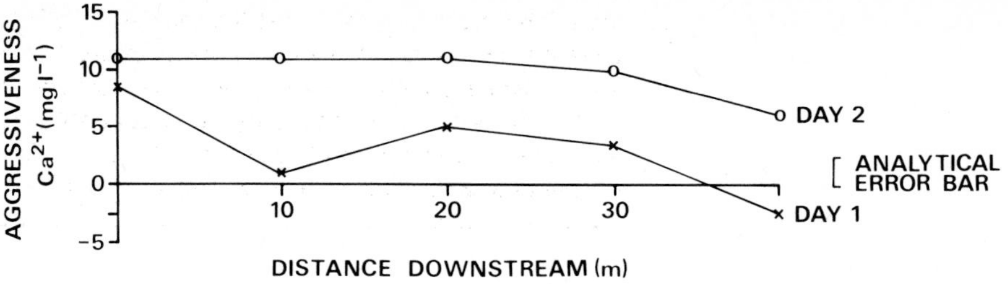

AGGRESSIVENESS Ca^{2+} (mg l^{-1})
15
10
5
0
-5
10
20
30
DAY 2
DAY 1
ANALYTICAL ERROR BAR
DISTANCE DOWNSTREAM (m)

Figure 3.14 Downstream changes in aggressiveness, Carboniferous Limestone, o——o Day 1, X——X Day 2 (After Trudgill, 1979)

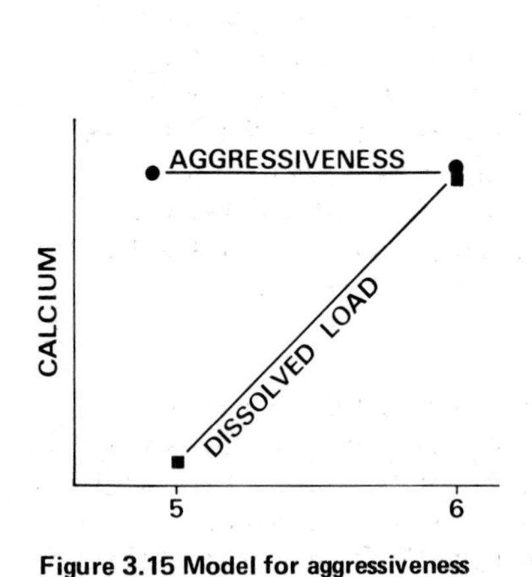

AGGRESSIVENESS
DISSOLVED LOAD
CALCIUM
5
6
RAINFALL
STREAM SWALLET
1
SOIL
5
2
6 DRIP
3
CAVE
RESURGENCE
4
AGGRESSIVENESS
CALCIUM
DISSOLVED LOAD
1
2
3
4

Figure 3.15 Model for aggressiveness studies in limestone drainage systems

variation occurs, it is clear that the general trend is for aggressiveness to decrease downstream from the peat bog—limestone junction. The trend of these data can be compared with downstream changes in stream-bed erosion rates described in section 4.2.

A general model for the study of limestone drainage systems is shown in ***Figure 3.15***. Aggressiveness can be studied at any accessible point in the system, but generally decreases as water travels through soil and bedrock and downstream in cave systems.

3.8 BIOLOGICAL PROCESSES

Some biological processes are relatively easy to observe, such as the mechanical action of tree roots in prising apart boulders. These commonly cited processes are, however, of far less importance than biological processes which take place in relation to the production of organic acids and carbon dioxide. Organic acids can dissociate to produce hydrogen ions which can become involved in hydrolysis (***see*** section 3.2). Organic acids can also directly incorporate some cations from minerals into their chemical structures – a process known as ***chelation.*** Additionally, chemical elements may be chelated from rock minerals by lichens growing on rock surfaces. Carbon dioxide produced by respiration of roots, animals and micro-organisms is a source of acidity in soils, as was shown in section 3.2. Field and laboratory experimentation can be undertaken to illustrate the actions of many of these processes and to test hypotheses about their effects, though obviously the actions of tree roots are difficult to determine experimentally! The study of biological processes involved in weathering is not a well developed subject and thus it is difficult to recommend simple, well-tried and tested methods. However, some of the important considerations are discussed below.

3.8.1 Organic acids

Organic acids, through their chelating ability, can mobilise chemical elements from minerals which are resistant to water charged with carbon dioxide. This applies especially to some of the silicate minerals. Huang and Kiang (1972) showed how salicylic acid and aspartic acid could dissolve more silica and aluminium from feldspars than could deionised water (***Figure 3.16***). Similarly, Atkinson and Wright (1967) demonstrated that an artificial chelate, EDTA, could mobilise far more iron and aluminium than could water

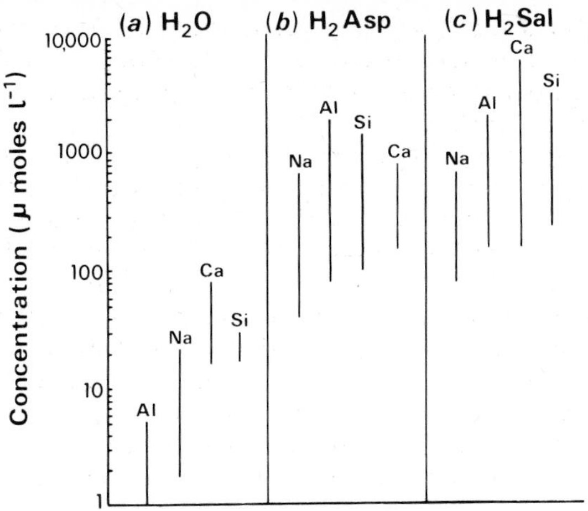

Figure 3.16 Dissolution of feldspar minerals in distilled water and two organic acids, aspartic and salicylic (After Huang and Kiang, 1972)

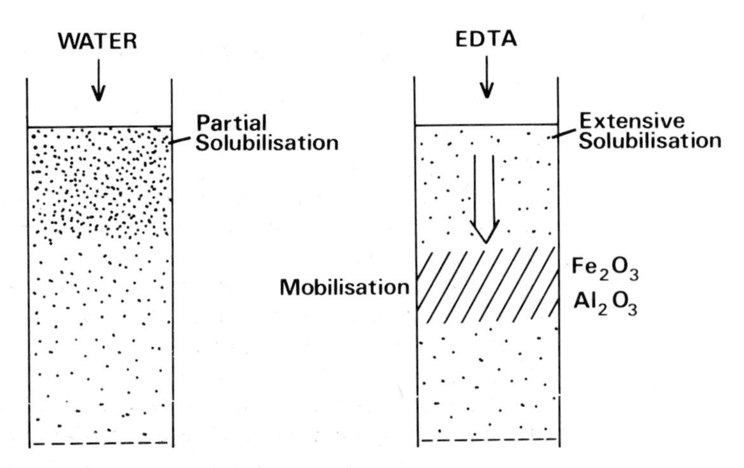

Figure 3.17 Chelation experiments (after Atkinson and Wright, 1967), showing mobilisation of iron and aluminium in soil columns using a chelate, EDTA

(*Figure 3.17*); Bloomfield (1953 a, b and 1954 a, b, c) obtained confirmatory evidence in a series of experiments using naturally derived leaf chelates. Replications of these experiments may be made if resources are available, but simple observations may be made as suggested below.

Organic acids can be used for experimentation in the same manner as described for inorganic acids in section 3.3. Some organic acids may be purchased in pure form from chemical suppliers and the more complex humus acids may be extracted from leaves, peat or other organic material. In the latter case, the material may be shaken with distilled water and the resulting solution filtered. The efficacy of organic acids to attack minerals may be demonstrated by the chemical analysis of test solutions before and after contact with minerals, as suggested in ***Table 3.9*** for limestone fragments. The difficulty with this work is that the concentrations of organic acids in soils and other natural situations are not easy to assess without advanced laboratory equipment, thus the data such as are shown in ***Table 3.9*** and further below, merely demonstrate that an effect may exist, but do not necessarily quantify the effect as it may occur in nature.

TABLE 3.9 ANALYSES OF 100 ml SOLUTIONS AFTER 24 h CONTACT TIME WITH LIMESTONE CHIPPINGS

Solution	$CaCO_3$ (mg ℓ^{-1})
Distilled water and carbon dioxide	300
0.01 M acids	
Oxalic	200
Acetic	1 000
Lactic	1 800
Salicylic	8 400
Citric	9 200
Tartaric	19 800

3.8.2 Leaf extracts

To test the hypothesis that acid leaf extracts can cause solubilisation of calcium carbonate.

Procedure

(1) Shake 10 g air-dried leaves for 1 h with 25 ml distilled water. Use several subsamples, with leaves from a wide range of trees, including separate samples of fresh and fallen leaves and three replicates of each subsample.
(2) Filter the solution.
(3) Measure the pH of the solution (***see*** p. 60) and titrate an aliquot for calcium – ***see*** p. 67 (the calcium values are liable to be low).
(4) Add 1 g powdered calcium carbonate.
(5) Shake for a further 1 h and filter again.
(6) Remeasure the pH and titrate a new aliquot for calcium.
(7) Note any changes in pH and calcium from (3) to (6). Plot any increases in pH or

calcium against the original pH of the solution. The data on calcium uptake can be plotted against original pH, taking the mean of three replicates as each data point, and plotting separate graphs for fresh and fallen leaves. If, say, 20 tree species were originally chosen, a regression analysis may be performed (Gregory, 1963, p. 207) or the correlation between pH and calcium uptake may be assessed using a test such as Spearman rank correlation (Briggs, 1977a, pp. 87–94). If the experimental design was designed to test differences between two groups of trees, say confiers and broad-leaves, then a significance test should be applied to the data groups, such as student's 't' test, or if the data are ranked, the Mann–Whitney U test (Briggs, 1977a, pp. 91–93 and pp. 87–94).

Problems

In general, increased amounts of calcium should be found in extracts of higher acidity (low pH). This assumes that the organic acids are ionising in water to produce hydrogen ions, but this may not always be the case. If chelates are present, however, these may be only weak acids in terms of dissociation of hydrogen ions, but nevertheless able to chelate large amounts of calcium. Lichen acids are often in this category (Hale, 1967, p. 89). An additional problem is that calcium strongly held in organic chelates may not be measurable by EDTA titration, depending on whether EDTA is a stronger chelator than the organic acid. This chelatory effect will be less marked on more decomposed material where the organic matter is broken down into the more simple acids.

Results

The potential of each leaf material examined to solubilise calcium can be discussed and the effectiveness of leaf leachates in nature for example, in rain dripping from leaves or percolating through litter can be discussed for each tree.

3.8.3 Bark extracts and stem flow

Water flowing down the bark of trees commonly becomes acid as it picks up organic acids, for example, gallic acid, a tannin derivative. Two useful measurements may be made, firstly, using small flakes of bark (taking care not to disfigure a living tree) and repeating the above experiment, and secondly by actually collecting stemflow water. A

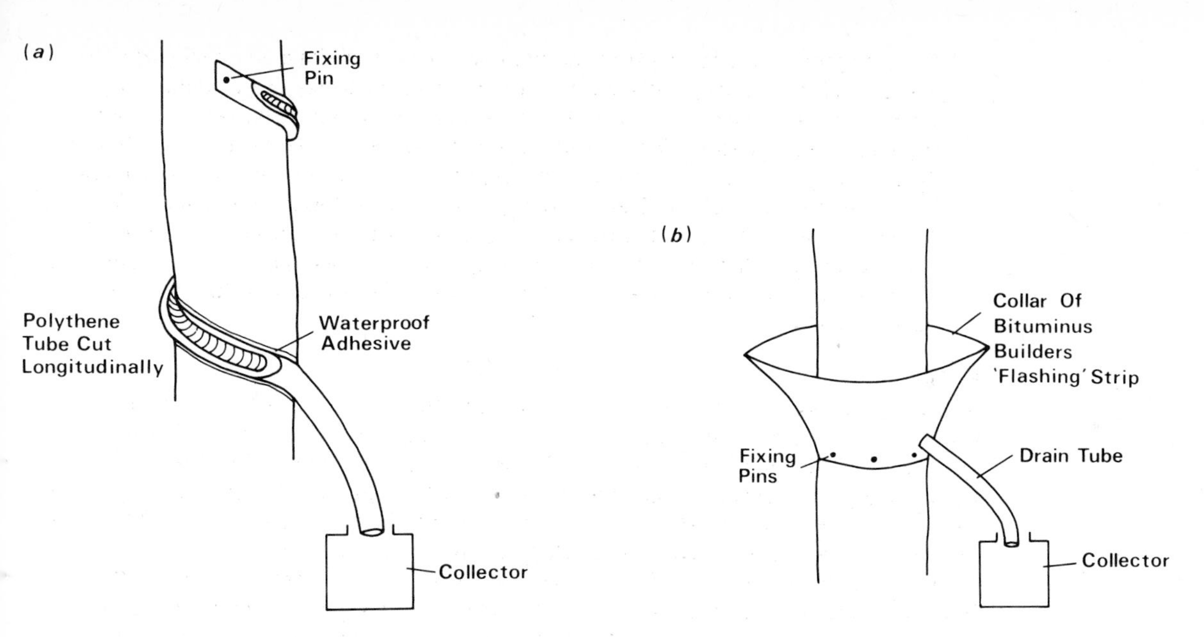

Figure 3.18 Stemflow collecting collars

stemflow collar can be constructed to collect water dripping down the bark (***Figure 3.18***). It is important to obtain a good seal between the collar and the bark using a proprietary builders' sealant, especially on rough-barked trees. The stemflow collected in plastic containers can be treated as the solutions above, using steps (3) to (7). Stemflow acidity values commonly fall as low as pH 3. Some sample values are shown in ***Table 3.10***.

3.8.4 Mobilisation of iron

Even if analytical methods are not available to replicate the measurements of Atkinson and Wright and Bloomfield (***see*** p. 85), it is possible to demonstrate the mobilisation of

TABLE 3.10 UPTAKE OF CALCIUM CARBONATE IN LEAF AND BARK EXTRACTS, MALHAM TARN, UK

Tree	Material	pH(1)	pH(2)	$CaCO_3$ (1) (mg ℓ^{-1})	$CaCO_3$ (2) (mg ℓ^{-1})
Larch (*Larix* sp.)	leaf	5.25	4.90	143	128
	bark	4.15	3.70	65	75
Yew (*Taxus baccata*)	leaf	6.00	6.15	53	60
	bark	4.35	4.00	35	95
Sycamore (*Acer pseudoplatanus*)	leaf	5.30	4.90	295	270
	bark	5.60	5.55	38	120

1 g bark or leaf material shaken with 20 ml distilled water
pH(1) = 5–10 min contact time
pH(2) = 1 h contact time
$CaCO_3$ (1) = extra saturated with 'Analar' calcium carbonate
$CaCO_3$ (2) = extract in contact with Carboniferous Limestone rock tablet for 24 h

iron. The experiment is a long-term one and should be run for several months. It consists of four boxes with drain holes at the base and glass fronts for observation. Alternatively a perspex tube with a basal nylon net may be used. Each is filled with a well-mixed sand from an inland source, with a tissue or filter paper placed on top. In one box, a layer of pine needles or other confier needles should be added on top of the tissue or filter paper. In the second box, a layer of deciduous leaves may be placed on top of the tissue or filter paper. A third box should have no leaves placed on the tissue or filter paper, to serve as a control. To each of these boxes, distilled water should be added, equal quantities to each box, applied at the same times; 500 ml per week per box should be an adequate application for a 20 cm X 20 cm X 30 cm high box. To the fourth box equivalent amounts of the artificial chelate EDTA should be added (as 0.025M, 9.306 g ℓ^{-1} ethylene-diamine tetraacetic acid, in the disodium salt form). If facilities for iron analysis exist then the effluent can be monitored for iron content and at the end of the experiment the soil may be analysed for iron distribution in the vertical profile. However, in the absence of such facilities, brown iron staining may become visible over time towards the base of the box as iron is mobilised from the sand and carried downwards by artificial or natural chelates. Observation can test the effects of different leaf litter types and the effects of water and EDTA. Naturally, no mobilisation will take place if there is no iron in the sand, so unweathered sand provides the best material for use. In addition, care must be taken to apply enough water to ensure a downward movement of water especially in large columns or in warm laboratories where evaporation will be high.

3.8.5 Carbon dioxide in soils

Carbon dioxide is a major source of acidity in soil water. For this reason, carbon dioxide content is an important property to measure, but it is often easier – and more relevant – to measure the acidity of the soil solution directly. However, gas-sensitive probes are available which can be used to measure soil carbon dioxide and thus the relation between acidity and carbon-dioxide content may be studied. If a soil is acid, but has a low carbon-dioxide productivity, then the source of the acidity is most likely to be organic acids or a chemical reaction, such as the oxidation of pyrite (Vear and Curtis, 1981). High carbon-dioxide content may not necessarily mean that there is a high potential for weathering, however, since gaseous build-up in soil may mean that the soil is impermeable; thus weathering products may not be removed readily. A gas-sensitive probe is illustrated in ***Figure 3.19*** (***see*** Appendix 1 for source). Soil air is drawn through a detector tube. A probe sheath is necessary to prevent the tube entrance from becoming blocked with soil, as shown. The tube should be inserted into the soil and the edges of the hole sealed to prevent atmospheric air from entering and diluting the soil air. Results for inorganic and organic soils are shown in ***Table 3.11***. Equilibration of carbon dioxide with calcium carbonate can be assessed from ***Figure 3.20***, though this describes the theoretical amount of calcium carbonate that could equilibrate with given amounts of soil carbon dioxide. Actual amounts of calcium carbonate in solution in runoff from soil-covered areas may be greater than this theoretical curve might predict if other sources of acids are present, or less if the soil is impermeable. For most studies of weathering under soil, the pH of the soil, the pH of percolating soil solutions and some measure of soil permeability (e.g. Briggs, 1977b, p. 83) are probably the most useful factors to study.

TABLE 3.11 CARBON-DIOXIDE LEVELS MEASURED IN INORGANIC AND ORGANIC SOILS (AFTER SMITH AND ATKINSON, 1976)

Soil type	Carbon dioxide (%)
Peat soils	0.5–1.5
Leaf litter	0.5–1.8
Calcareous soils	0.2–2.0
Acid mineral soils	0.2–1.6
Waterlogged soils	0.5–3.0

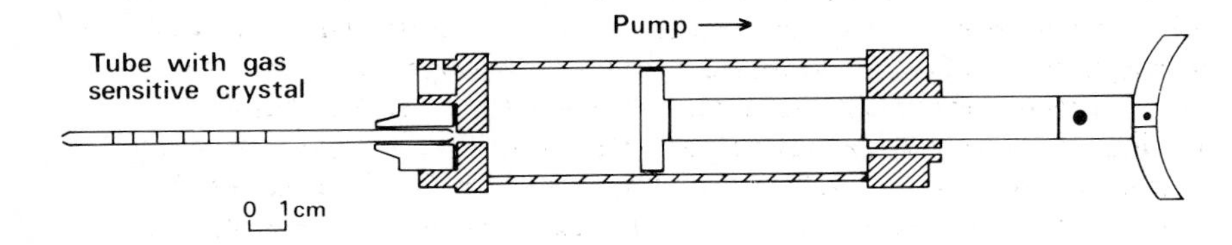

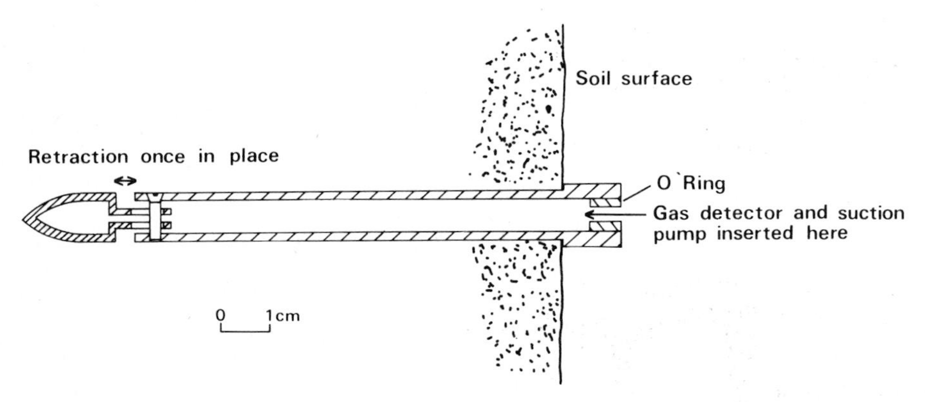

Figure 3.19 Soil carbon-dioxide measurement apparatus (Gastec probe, *see* Appendix 1)

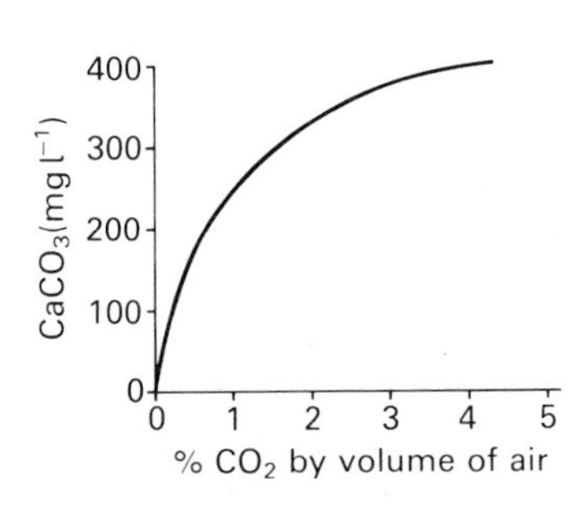

Figure 3.20 Equilibration of calcium carbonate in solution with carbon dioxide (From Smith and Atkinson, 1976)

3.8.6 Lichens and algae

Lichens and algae can colonise most bare rock surfaces and are often the first sources of weathering potential, mainly by the production of chelates. Colonisation can be demonstrated by leaving fresh-cut stone samples in the open, and examining them each month with a powerful hand lens or direct viewing microscope. Alternatively, lichen colonisation can be studied on a sequence of dated gravestones, provided, of course, that permission has first been sought from church authorities. Jones (1965) showed how lichens may penetrate into limestone along calcite cleavage planes (*Figures 3.21, 3.22*). In general softer, more soluble rocks which have well-defined crystal cleavage planes are more prone to lichen penetration than harder rocks. The action has the effect of loosening crystals and increasing the surface area open to water contact after the decay of the lichen, though, during growth, calcium carbonate precipitation may occur on the surface of some lichens, leading to an increase in surface hardness.

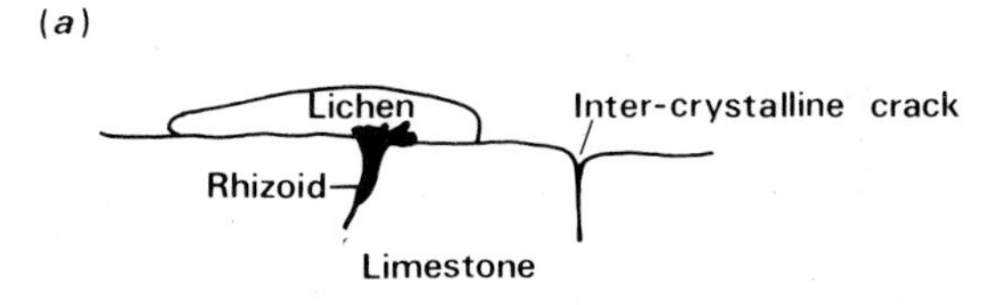

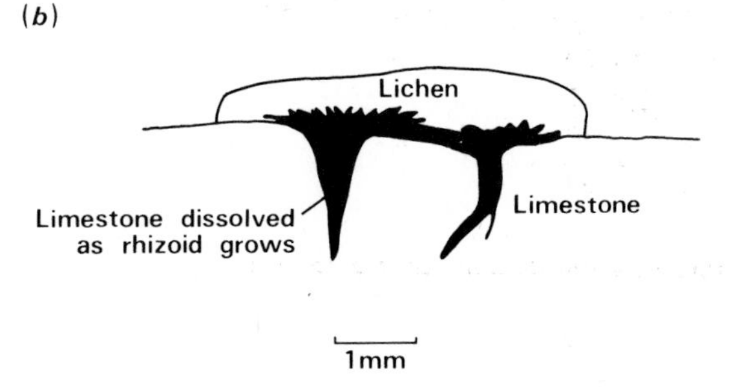

Figure 3.21 Penetration of lichen surface by lichen rhizoids (After Curtis, Courtney and Trudgill, 1976)

Figure 3.22 Thin section of limestones showing lichen rhizoids *in situ* (After Jones, 1965)

3.8.7 Summary

As emphasised initially in this section, biological processes may be difficult to observe experimentally and to measure quantitatively as it is often difficult to replicate natural conditions in the laboratory. However, some useful observations may be made, most of them providing adequate demonstrations of the operations of some of the processes involved.

3.9 WEATHERING PROFILES

In the sides of road cuttings or quarries, vertical sections can often be seen where the top portion, nearest the surface, is very different from the section below. In some cases this is because different material has been deposited on top of a bedrock, or because rocks have been laid down in contrasting strata. However, in situations where the bedrock is homogeneous vertically and where no material has been deposited on top of it, a weathering profile may be in evidence. Frequently, the most easily soluble material has been removed from the surface and this may have been redeposited further down the profile; in addition, alteration and reformation of minerals may have occurred.

Sampling of the unweathered rock, and comparison with overlying ***in situ*** weathered material, frequently reveals that some minerals have been lost and new ones formed during weathering. This may be evident from visual observation; for example pyrite cubes may be visible as gold-coloured cubes in unweathered shales in the lower portion of a section, but above this they may have weathered out to leave cuboid holes (***Figure 3.23***). In other cases, mineral analysis by X-ray diffraction (XRD) can reveal the changes. During XRD X-rays are passed through a mineral powder and each crystal present in the powder diffracts the rays in a characteristic way, the presence of combinations of rays of different

Figure 3.23 Box-work lattice produced by weathering out of pyrite cubes (Photograph: J. Owen)

wavelengths permitting the identification of the mineral present. A record of the diffracted rays present is produced on a chart by a detector. The system is shown schematically in ***Figure 3.24*** and an example is shown in ***Figure 3.25***. The example shows that chlorite has been lost during weathering as the chlorite peaks are present in the bedrock but are absent in the soil.

Calcium carbonate is often readily soluble in earth surface conditions and the study of weathering profiles frequently reveals it to be lacking from the upper portion of a section. The depth to which it has been lost varies with the permeability of the material, the rainfall of the area, the original carbonate content and the time elapsed since the deposition of the carbonate material. The effect may be seen in glacial deposits; older tills frequently contain less calcium carbonate than younger tills. These profiles can be investigated by the use of sampling and chemical analysis, or more readily using field acid bottles (see Briggs, 1977a, pp. 34–35; 1977b, pp. 114–117). An example of till-profile acid reactions

TABLE 3.12 ACID REACTIONS IN A TILL WEATHERING PROFILE, CO. CLARE, EIRE

Depth	$CaCO_3$ (%)	pH	Reaction with 10% HCl
Calcareous till with high clay content			
0	5	7.0	weak positive
10	8	7.0	strong
20	10	7.2	strong
30	50	8.5	violent
40	60	8.8	violent
50	65	8.8	violent
60	70	8.9	violent
Leached till with high sand content			
0	0	5.5	none
10	0	6.0	none
20	0	6.5	none
30	2	6.5	weak positive
40	10	7.0	strong

is shown in ***Table 3.12. Figure 3.26*** shows the losses of calcium carbonate from a soil profile described by Birkeland (1974). After approximately 10 000 years the initial, uniform distribution is altered so that no calcium carbonate is present down to 50 cm. This is a reasonably clear example of a weathering profile since independent evidence, largely based on soil texture, indicates an initial uniformity. The situation should not be confused with one where the initial distribution already displayed a pattern of increase down the profile (***Figure 3.27***).

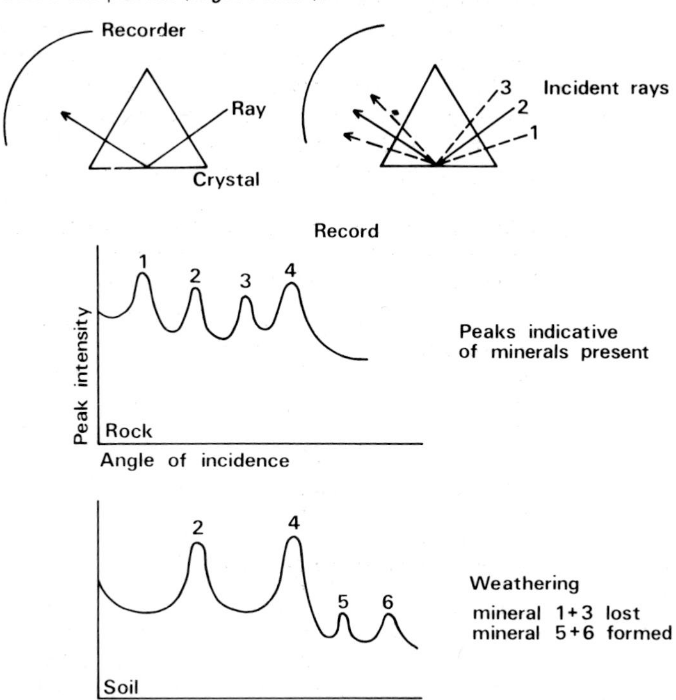

Figure 3.24 X-ray diffraction system; combinations of peaks are used to identify minerals

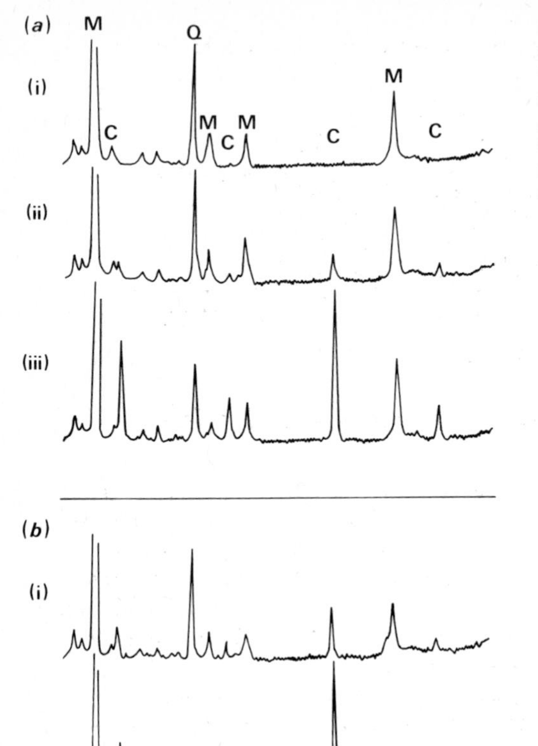

Figure 3.25 Loss of chlorite during weathering. X-ray diffraction profiles from Silurian mudstone in central Wales. M = mica, C = chlorite, Q = quartz (After Adams, Evans and Abdulla, 1971)
(*a*) podsol soil: (i) A horizon, (ii) B horizon, (iii) C horizon. (*b*) brown earth: (i) B horizon, (ii) C horizon

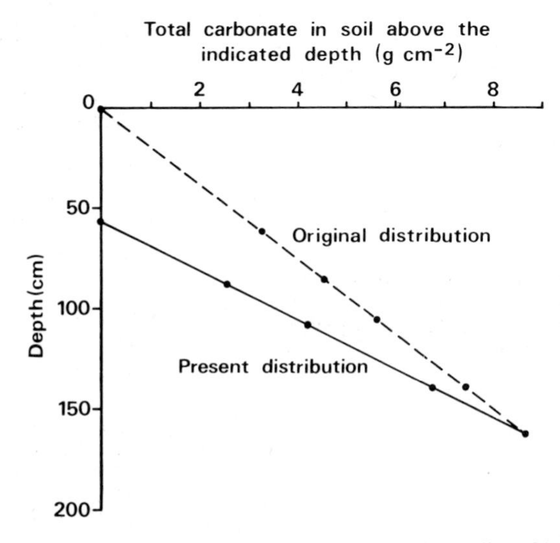

Figure 3.26 Leaching of calcium carbonate in a soil profile (After Birkeland, 1974)

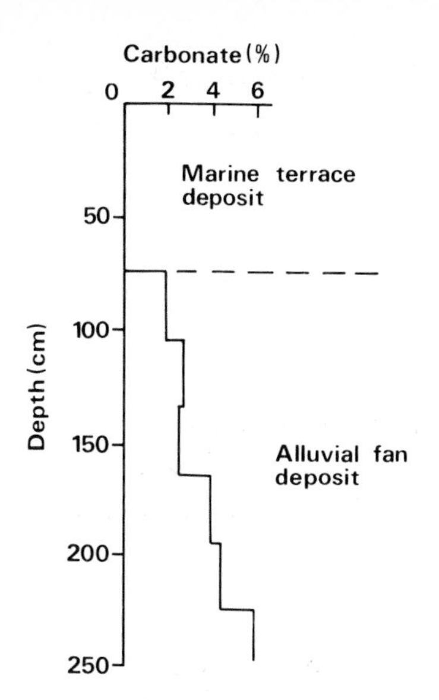

Figure 3.27 Differential distribution of calcium carbonate in a soil profile due to differences in deposition (After Birkeland, 1974)

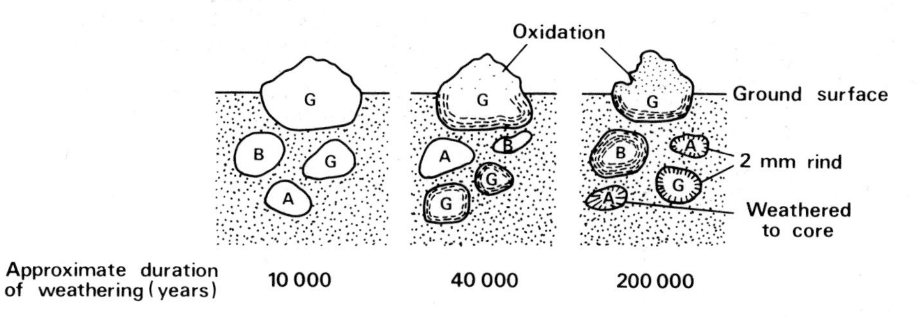

Figure 3.28 Weathering of boulders with length of time exposed (After Birkeland, 1974) G = granite, A = andesite, B = basalt

Data from the study of weathering profiles can normally be interpreted in the light of the processes outlined in ***Figure 3.4a*** (p. 54), with removal of soluble material from the top and redeposition below. In addition, weathering crusts, that is, an outer layer of altered material, can often be seen in relation to weathering profiles, with rocks nearer the surface or those most exposed to weathering showing the most well-developed crusts. An example of weathering through time is shown in ***Figure 3.28***. In 200 000 years a weathering crust or rind about 2 mm thick has developed.

Analytical work on weathering profiles is normally limited to field observation of visible changes in colour and mineralogy, complemented by sampling and chemical, textural and mineralogical analyses in the laboratory. Samples may be analysed as discussed by Briggs (1977 a, b). Vertical changes can then be plotted graphically, and discussed in the light of the likely processes in operation in the past and at present. However, experimentation of the type outlined in section 6.6 (p. 161) may be useful where the most soluble chemical elements are being studied. Otherwise, experimentation is difficult since, as indicated by the time scales mentioned above with reference to published studies, weathering profiles normally only evolve over several hundreds of years.

PART III

EROSION

CHAPTER 4 SURFACE LOWERING OF HARD ROCKS

4.1 Introduction

Erosion involves the removal and transport of material from its original site by an agency such as running water, moving ice or wind. Erosion may be enhanced by prior weathering, but weathering is not a prerequisite for erosion. A knowledge of erosion rates can be used to understand the differential evolution of relief and is fundamental to the understanding of landform evolution. The significance of measuring erosion rates lies in the fact that it is important to understand the spatial distribution of erosion rates in order to be able to understand how the landscape evolves. It may be possible, for example, to ascertain that two rocks which are juxtaposed have different erosion rates, one eroding faster than the other because it is softer; or that identical rocks could be eroding differentially because erosion agents are more powerful in one place than in another. Such measurements can show why one part of the landscape may be more upstanding than another and thus provide explanations of erosional landforms. Measurements of erosion rates therefore assist in the understanding of the relationships between processes and landforms and are fundamental to geomorphological study. It is, however, important to remember that if ***current*** erosion rates are measured, they may not always be of direct use in the explanation of erosional landforms. This is because most landforms have evolved over long time spans and many show features inherited from past regimes when erosional processes and rates may have been very different to those operating at the present day.

Rates of erosion may be assessed directly by repeated measurements of surface lowering if the current rates are fast enough and if the techniques used are accurate enough. Alternatively, they may be assessed indirectly by comparisons of relief between areas which have suffered erosion and areas which have not. In addition to measurements of surface lowering, measurements of mass or volume loss can also sometimes be made. Measurements made at the present day cannot confidently be extrapolated back through time, as explained above. However, they can be compared with measurements gained on a long-term basis. These may be obtained if the age of an original surface is known, as would be the case, for example, with a glacial planation surface where the date of retreat has been ascertained by ^{14}C dating of glacial deposits. The amount of surface lowering may be measured by comparing the level of an eroded surface with that of a preserved glacial surface (***see*** section 4.7). Experimental work can also be undertaken, with observations on test samples of rocks in the field and on comparisons of rock erosion under control laboratory conditions. Several examples of direct and indirect measurements of erosion are described below, together with some descriptions of experimental work.

4.2 MICRO-EROSION MEASUREMENTS

Many erosion rates are very slow when compared with the time span of human life available to measure them. This means that if measurements are to be made of erosion rates which are operating at the present day, then extremely accurate methods have to be used to obtain reliable, interpretable results in a few years. One such method involves the use of a ***micro-erosion meter.*** This consists of a micrometer which measures the height of a rock surface relative to a reference position.

4.2.1 Method

The micrometer measurement is made with a dial gauge and a micrometer probe which is mounted onto a tripod framework (***Figure 4.1***). The tripod arrangement gives the instrument stability and it rests on three reference studs permanently inserted into the rock. The height of the surface is measured relative to the studs and the reading of this height can be repeated at intervals, usually yearly, though more frequent measurements may be made on rapidly eroding rocks. In this way the rate of lowering of the surface relative to the studs can be measured. The procedure is illustrated in ***Figure 4.2***.

A number of precautions must be taken when using a micro-erosion meter. It is important to protect the studs in the period between measurements so that any surface-lowering recorded is due to rock lowering and not to movement of the studs. A firm seating for the studs is provided by 'Rawlbolts' manufactured for holding bolts in masonry. The upper end of the studs themselves should be hemispherical. They can be made by welding stainless-steel ball bearings onto bolt heads or by turning bolts or threaded rods on a lathe. Ball bearings can also be affixed to countersunk bolt heads by

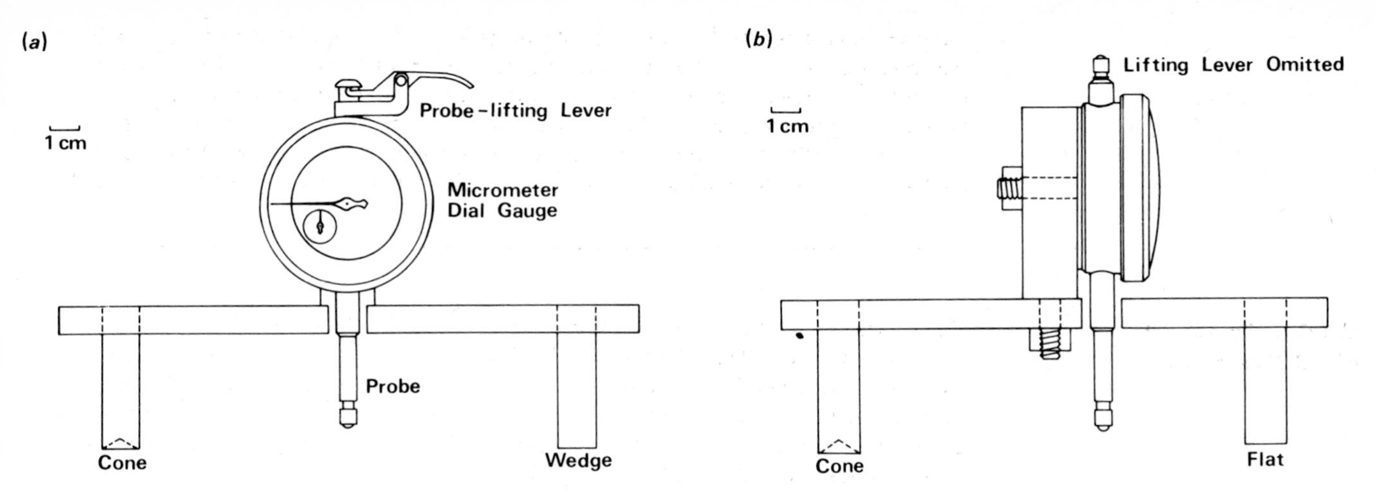

Figure 4.1 Micro-erosion meter: (a) face view; (b) side view (From High and Hanna, 1970)

the use of a fixative such as 'Araldite' but in very vigorous or corrosive environments (such as under sand abrasion in sea water) the balls may become removed unless given very good protection with plastic caps. Studs and stud protection are shown in ***Figure 4.3***.

The meter tripod legs are machined so as to be stable and precisely located; thus cone, wedge and flat shapes are used, machined into the legs, which are preferably made from alumina. Stainless-steel legs may wear considerably with frequent use. Technical specifications for the meter are given by High and Hanna (1970) and by Trudgill, High and Hanna (1981). The latter reference gives details of a traversing micro-erosion meter (***Figure 4.4***) where the micrometer dial gauge is mounted on a movable block. The arms of the block are located on large ball bearings and, by turning the block around within the base, and also by turning the base around, several readings can be taken at one site. This permits the evolution of micro-topography to be studied.

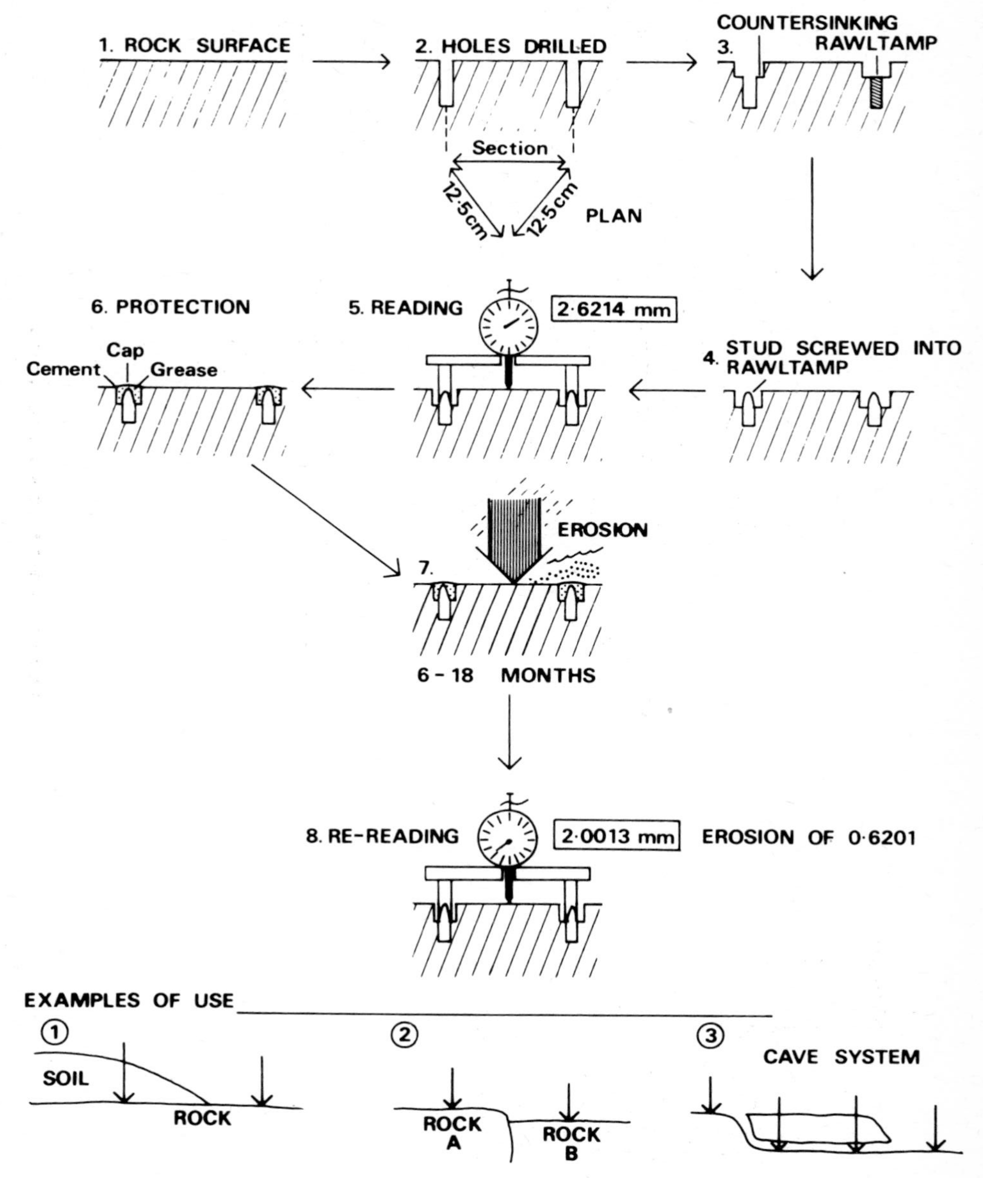

Figure 4.2 Micro-erosion measurement procedure

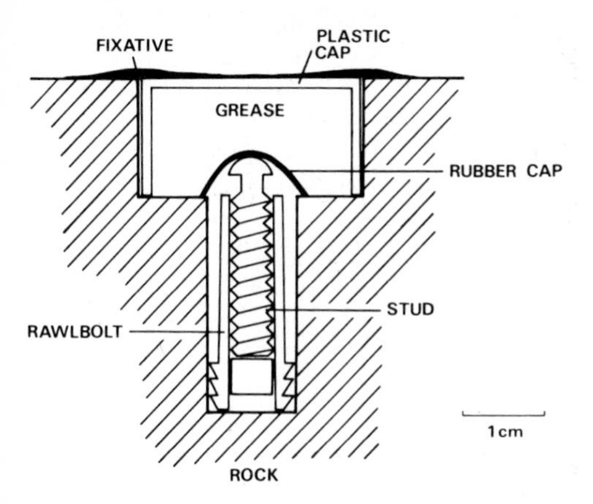

Figure 4.3 Studs and stud protection

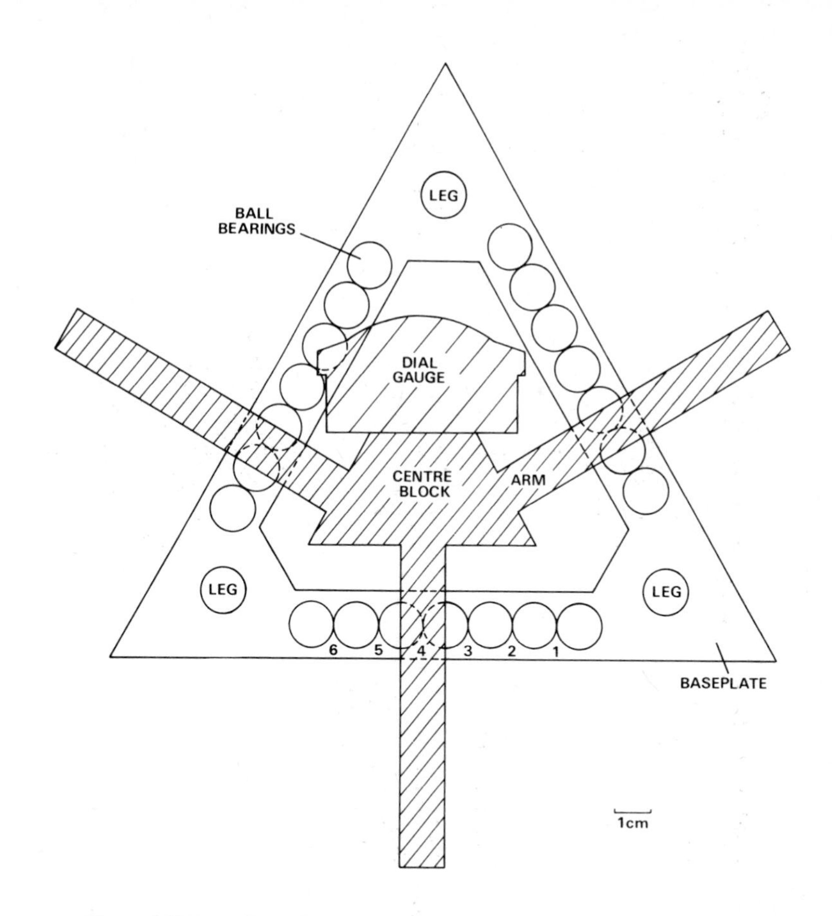

Figure 4.4 Traversing micro-erosion meter

4.2.2 Discussion

The micro-erosion meter has been widely applied to the study of limestone erosion. Many hard limestones present a sound surface upon which it is easy to make valid measurements without the surface crumbling. Limestone erosion rates are of the order of 0.01 mm per year (mm a^{-1}) and above, and these are rapid enough to be measured over intervals of a few months or years. Other rocks may also be used, but if the erosion is taking place by granular disintegration or by small blocks becoming detached, then very erratic readings may occur.

Figure 4.5 illustrates the rate of lowering for one cave stream-bed site in Co. Clare, Eire. Results for three sites on the limestone surface are shown, taken at five intervals over 500 days. The data plot on straight lines showing that short-term measurements over 50–100 days give results which are also valid in the long term. The actual erosion rates per year are shown in ***Table 4.1.*** They decrease from 0.4–0.5 mm a^{-1} nearer the swallet to 0.05 mm a^{-1} at the outflow of the cave system where the water is less aggressive (aggressiveness is discussed in section 3.7).

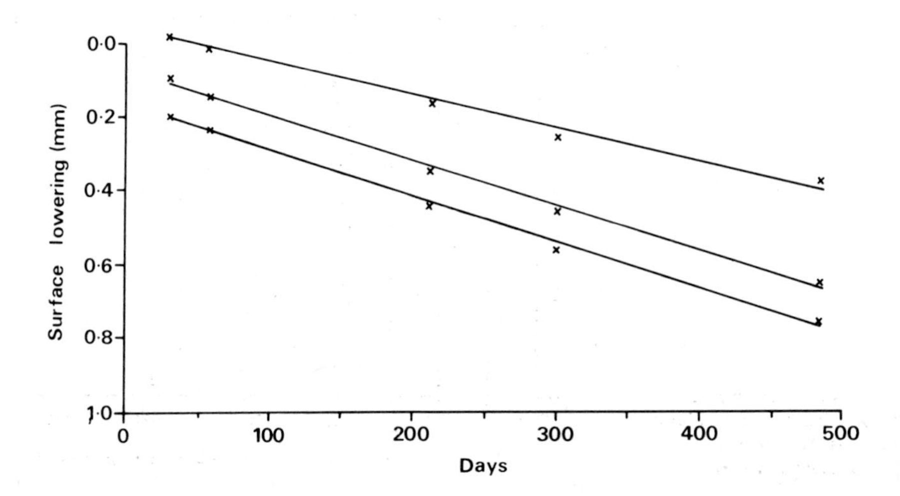

Figure 4.5 Progression of surface lowering in cave stream bed measured by micro-erosion meter (After High and Hanna, 1970)

TABLE 4.1 AVERAGE EROSION RATES FOR THREE SITES AT DIFFERENT LOCATIONS IN A STREAM (FROM HIGH AND HANNA, 1970)

Location	Site	Aggressivity of stream to $CaCO_3$	Distance from beginning of stream (m)	Mean erosion rate (mm a^{-1})
Stream at swallet	Stream bed	Very aggressive	365	0.5
Stream in cave	Stream bed	Aggressive	820	0.4
Stream at rising	Stream bed	Nearly saturated	2560	0.05

The frequency distribution of erosion rates from an intertidal limestone site is shown in ***Figure 4.6***. The mean rate is 0.2 mm a^{-1} but the range from 0.05 to 0.75 over the area of the measurement site indicates that a dissected micro-topography is evolving.

Data from Kirk (1977) for intertidal micro-erosion on different lithologies are shown in ***Table 4.2.*** Although the rates differ from year to year it is apparent that the mudstone studied is being eroded at a mean rate over twice the mean rate of erosion of the limestone.

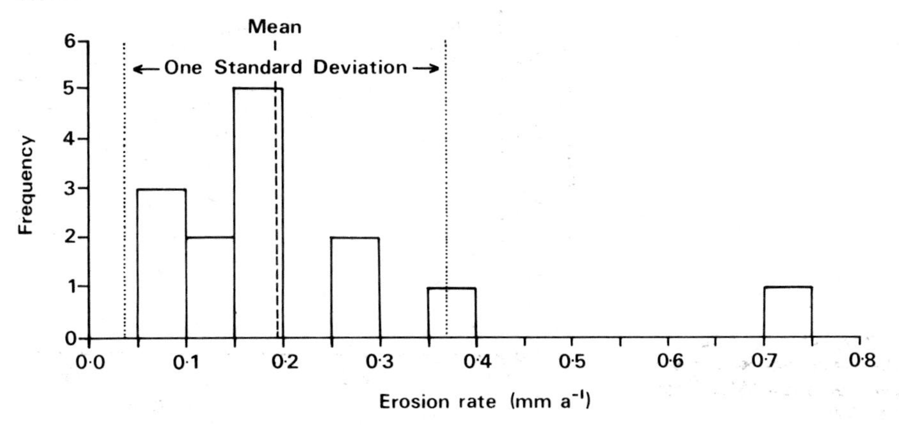

Figure 4.6 Traversing micro-erosion meter results, intertidally exposed limestone site, Co. Clare, Eire (From Trudgill, High and Hanna, 1981)

TABLE 4.2 MICRO-EROSION MEASUREMENT AND LITHOLOGY, KAIKOURA PENINSULA, S. ISLAND, NEW ZEALAND (AFTER KIRK, 1977)

Site lithology	Erosion rates ($mm\ a^{-1}$) 1973–4	1974–5	Mean
1. Limestone	0.10	0.67	0.38
2. Limestone	0.99	1.29	1.14
3. Mudstone	0.96	1.96	1.46
4. Mudstone	2.32	2.68	2.50
5. Mudstone	0.82	0.46	0.64
6. Mudstone	2.49	2.32	2.41

Mean limestone: 0.76 $mm\ a^{-1}$
Mean mudstone: 1.75 $mm\ a^{-1}$

The role of individual erosion agencies can be assessed by designing a measurement network of sites where erosion agencies vary but where lithology and all other factors are constant. For example Robinson (quoted in Kirk, 1977) measured rates of erosion on one outcrop of mudstone on the North Yorkshire coast of England. Where abrasion by gravels occurred, the range of rates was 15.0–35.0 $mm\ a^{-1}$, but where gravel was absent the rate dropped to 0.1–1.0 $mm\ a^{-1}$. Similarly, Trudgill (1976b) measured rates of erosion of intertidally exposed coral limestone on Aldabra Atoll, Indian Ocean. On sheltered sites, rates measured were in the range of 1 $mm\ a^{-1}$ to 1.5 $mm\ a^{-1}$ but on coasts exposed to the dominant south-east trade winds the range rose to 3.0–4.0 $mm\ a^{-1}$. The higher rates recorded were also those were abrasive sand was present.

4.3 MASS-LOSS TABLETS

4.3.1 Method

Mass-loss tablets have been used as a measure of weathering potential, as described in section 3.6. The results can also be used to calculate an erosion rate, however, even though the procedure may not be wholly defensible. The data on mass loss are converted to a value for volume loss using the formula:

$$\text{density} = \frac{\text{mass}}{\text{volume}}$$

$$\text{i.e. volume loss } (v) = \frac{\text{mass loss } (m)}{\text{density } (d)}$$

This assumes knowledge of the density of the rock. Density can be calculated at the outset of the experiment if the volume and mass of a rock tablet are known. Volume loss can be expressed in millimetres of surface lowering during the time of emplacement in the environment by distributing the volume loss over the surface area of the tablet. The calculations are as follows:

$$\text{volume loss a}^{-1} = v = \frac{m}{d}$$

where v is in cm^3, m is in g, d is in g cm^{-3}

$$v = \text{height} \times \text{length} \times \text{breadth (or for a cylinder, } \pi r^2 h)$$

$$\text{thus } h = \frac{v}{a} \text{ where } a = \text{surface area } (2\pi rh + 2\pi r^2 \text{ for a cylinder})$$

Thus, knowing the surface area (a), and volume loss (v), an equivalent surface-lowering loss (h) can be calculated (Trudgill, 1975).

4.3.2 Discussion

The criticism of this method is that an equivalent lowering rate is calculated, not an actual rate. It also assumes that mass loss is equally distributed over the surface of the tablet. It is a method which can be used to derive estimates of erosion rates to compare with those gained by other means, and it is preferable to use some other more direct method if it is available.

4.4 GRAVESTONES AND DATED BUILDINGS

If the age at which a stone surface was originally exposed is known, then the amount of alteration, erosion and weathering which has taken place to the present day can often be assessed. For instance, the state of gravestones of known age can be compared (if permission has first been gained to work in a graveyard). Simple visual comparison may be made of similar stones of different ages and observations can also be made of the loss of definition of any inscription. The date of erection of many buildings can also be ascertained and the state of the stone observed; this approach is difficult, however, in that it can only employ subjective visual assessment; if a building has been cleaned then a fresh surface will have been formed since the date of erection.

A more specific example is provided by studies of St Paul's Cathedral, London, UK (*Figure 4.7*). Here, the Portland Stone balcony was finished in 1720, with a lead flashing

Figure 4.7 Weathering of Portland Stone on St. Paul's Cathedral, London; on building the lead filler between the blocks and the stone surfaces were level (Photograph: S.T. Trudgill)

between the stones. The stone has been eroded by dissolution in the rain, probably aided by acidity from air pollution. The amount of lowering of the stone relative to the flashing has been about 1.5–2.0 cm during the time since the building was erected. This is equivalent to an erosion rate in the range of 0.06 to 0.08 mm a^{-1} (Sharp ***et al***., 1982).

Emery (1941) used the rate of obliteration of dated inscriptions in a sandstone exposed intertidally at La Jolla, California, USA, to deduce that an erosion rate of 0.3 mm a^{-1} was occurring.

Such data have limited application in the context of landform evolution, but the susceptibility to weathering of contrasting rock types used as building stone can often be assessed using this approach.

4.5 EROSION PINS

Erosion pins are metal rods driven into rocks and they act as reference points to which rock surface retreat can be compared. They are of limited use in hard rock because the erosion rates are so small compared with those in softer sediments and soils (section 5.6). However, masonry pins may be driven into softer rocks where there are rapid rates of erosion. One successful application of this technique is that by Hodgkin (1964) who inserted steel pegs into an intertidal limestone notch exposed to marine erosion. He remeasured the sites after nine years and the results are shown in ***Figure 4.8.*** The rates of erosion detected were low and a period of 5–10 years is clearly necessary in this type of situation to measure the rate accurately. The rates Hodgkin measured are equal to a range of 0.22 to 1.0 mm a^{-1}, the highest rate occurring in the back of the intertidal notch where intertidal erosion processes are evidently concentrated. Here the data are of use in indicating the relationships between the concentrations of erosion processes and the evolution of morphology.

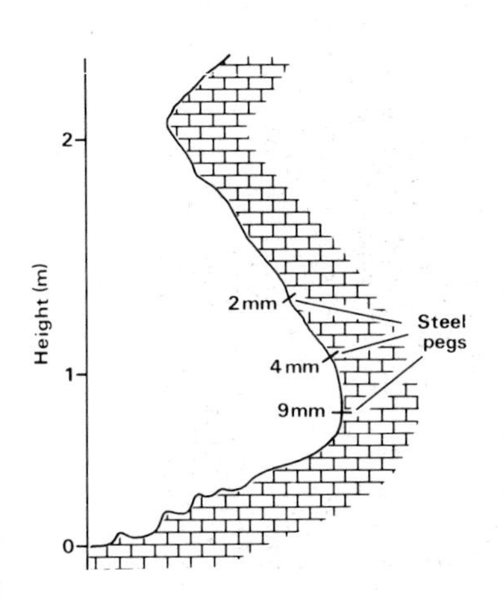

Figure 4.8 Erosion of intertidally exposed limestone measured by the use of steel pins over nine years (After Hodgkin, 1964)

4.6 DIFFERENTIAL EROSION

The assessment of differential erosion involves the measurement of relative height of different rocks under the same conditions. The main use lies in the assessment of the resistance to erosion of different rock types. For example, sandstones, limestone and volcanic rocks can be compared in a mountain stream bed or the resistance of dykes relative to the surrounding country rock can be compared. For micro-measurement work, an automobile tyre-tread depth gauge can be useful for measuring depth of grooves in rock surfaces. Dahl (1967), working in Norway, assumed a known level post-glacial surface and measured micro-erosion since glaciation by studying the indentation of surfaces between resistant quartz veins using the device shown in ***Figure 4.9***. Erosion rates ranged from near zero to 20 mm post-glacially, increasing with the presence of wet moss vegetation at low altitude (***Figure 4.10***). This means that, in the area involved, only small-scale features – of the order of 20 mm high – could have been formed post-glacially; the larger-scale subaerial surface landforms must therefore be essentially fossil and largely inherited from the previous glacial regime.

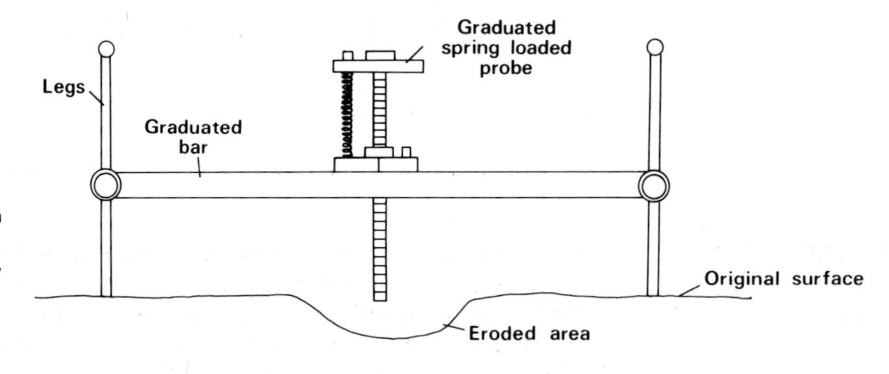

Figure 4.9 Dahl's apparatus for measuring micro-relief relative to resistant quartz veins. The apparatus is made of brass. The steel spring to the left of the pressure pin brings the pin back to the zero position after each reading. The pressure pin and the supporting legs are graduated in millimetres with a heavier mark for each fifth millimetre. The centre mark on the plate, which can be moved along the bar, can easily be set at the desired centimetre mark on the bar. During the measurement the position of the bar is stabilised by fixing the supporting legs with the aid of the side screws

Figure 4.10 Diagram showing the distribution of the mean values of micro-erosion in a granite area (N. Norway). Each mark represents 30–50 measurements (Dahl, 1967)
○ = mean value from localities with wet moss vegetation;
x = mean value from surface covered with snow for the greater part of the year;
✧ = mean value from block surface in a block-field zone;
● = mean value from other surface;
dots within dashed lines are from an area with marked glaciofluvial polishing

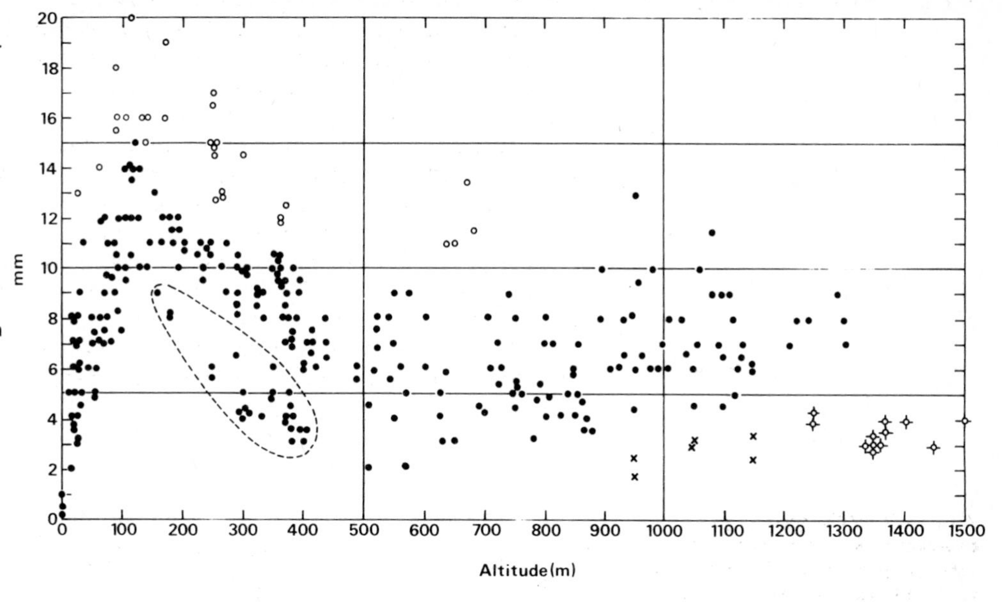

4.7 GLACIAL ERRATIC PEDESTALS

4.7.1 Method

When large glacial erratics are deposited on a fairly uniform, hard-rock surface, they commonly act to protect the surface beneath from subsequent weathering. Thus, over time, as the surrounding rock is eroded the rock beneath the erratic assumes the form of a pedestal. The difference in height between the pedestal and the adjacent rock can be used to estimate a rate of erosion since the time of deposition of the pedestal (*Figure 4.11*).

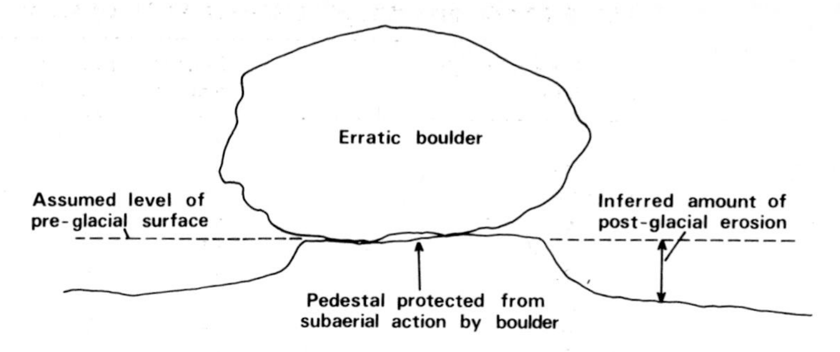

Figure 4.11 Glacial erratic pedestal

4.7.2 Discussion

This method involves several assumptions. Firstly, it is assumed that the erratic was deposited towards the end of the last glacial period and that subsequent erosion has taken place post-glacially. For much of the UK a time span of about 10 000 years is probably valid, but relics from previous glaciations may be found in areas not thought to have been covered in the last Glaciation (e.g. much of the Derbyshire limestone district). In addition, in Scotland and highland Wales, a more realistic deglacial date is nearer 8000 years ago. Secondly it is assumed that the boulder was not originally perched on a raised piece of rock during glacial deposition (nor was it embedded in till which has itself been eroded), and thirdly that the presence of the boulder has not led to greater runoff immediately around the boulder.

Table 4.3 lists some examples and calculated erosion rates for limestone glacial erratic boulders found on the west coast of Co. Clare, Eire. The method gives the observer an order of magnitude estimate of probable post-glacial rates of erosion. The use of such rates in interpreting the evolution of post-glacial landforms, however, must be carefully qualified. The erosion may not have been uniformly distributed throughout the time available. This may be tested to some extent by comparing the calculated annual rate of erosion with values gained from short-term methods (***see*** section 4.1). Thus, some indication may be obtained of the relation between present rates and past average rates. In addition, if it is desired to explain some feature other than that from which the rate was derived, it should be remembered that the erosion rates gained may not be transferable to other sites where conditions may have been different. Nevertheless, this

TABLE 4.3 EROSION RATES COMPUTED FROM GLACIAL ERRATIC PEDESTALS

Pedestal height (cm)	Erosion rate (assuming 10 000 years) (cm $a^{-1} \times 10^3$)	($mm\ a^{-1}$)
30	3.0	0.03
32	3.2	0.032
28	2.8	0.028
40	4.0	0.040
15	1.5	0.015
30	3.0	0.030
31	3.1	0.031
Mean 29	2.9	0.029

technique has yielded some useful indications of post-glacial erosion rates and has shown that in many cases post-glacial erosion has only modified a glacial landscape by a few tens of centimetres. Much of the landscape in these cases can thus be interpreted as essentially a fossil glacial one, little affected by current conditions.

4.8 MARINE BORERS AND GRAZERS

Several species of mollusc graze algae from rock surfaces, and remove some of the rock as they do so. Several bivalve molluscs such as ***Hiatella arctica*** also bore into rock substrata, especially into limestones. Others including ***Pholas*** spp. commonly bore into peats and other soft substrates (Evans, 1970).

4.8.1 Method

The depths of the holes bored can be measured using rods and, if the mollusc can be extracted without fracturing it (e.g. by chipping the surrounding rock away with a hammer and chisel) then it is possible in many cases to identify the major growth rings; these can frequently be taken to be annual, especially if the shell is stepped in long section. While there may be difficulty in precisely counting the rings, an estimate of boring rate may be gained by comparing the number of rings with the depth of the hole, assuming that the mollusc has been correctly identified and that it has actually produced

Figure 4.12 **Limestone bored by *Hiatella arctica* and also showing the finer borings produced by sponges (*Cliona*); scale bar is 1 cm (Photograph: J. Owen)**

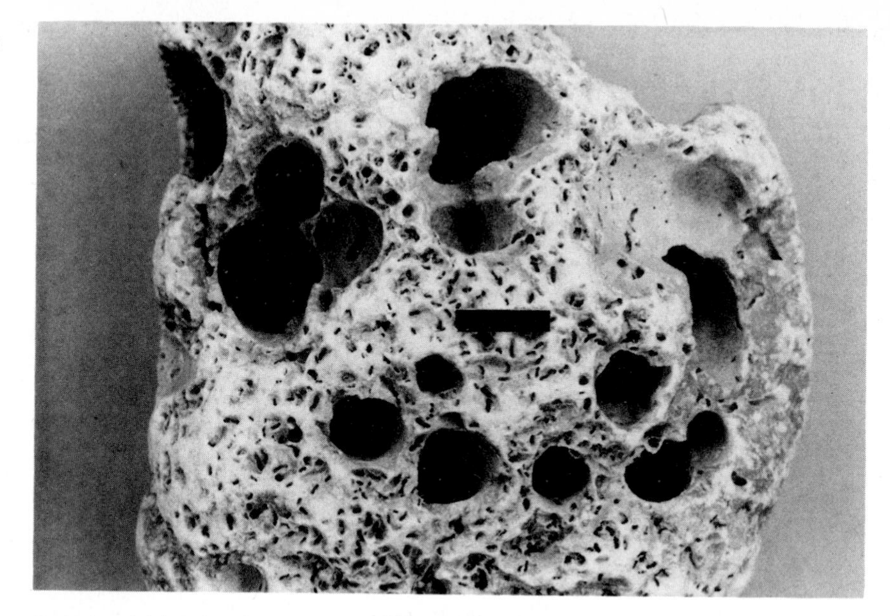

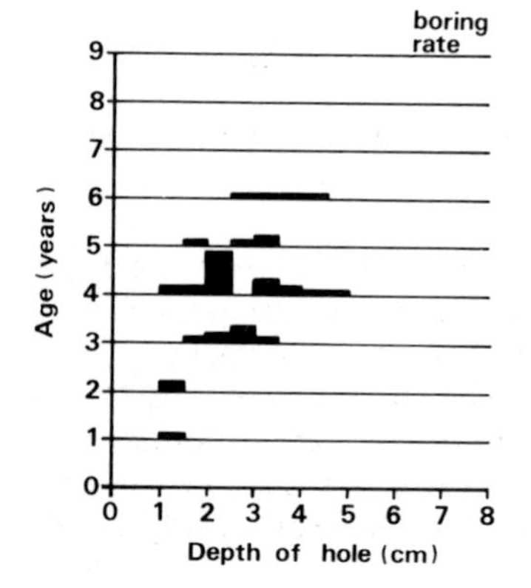

the hole which it is occupying. An example of a rock bored by ***Hiatella arctica*** is shown in *Figure 4.12*.

Sea urchins (echinoderms) also bore into rock substrates by means of teeth and spines. Chemical treatment of the test ('shell') can reveal annual growth rings and a boring rate can be calculated as for the molluscs. Details of the method are not given here for reasons of conservation but the results gained from a study of ***Paracentrotus lividus*** in Co. Clare, Eire are shown in ***Figure 4.13.*** There is a scatter of data for each age class but the depth of hole tends to increase with age, yielding a mean boring rate of around 1 cm a^{-1}. Other studies have used successive observations at one site or laboratory control investigations.

Figure 4.13 Boring rate of *Paracentrotus lividus*

4.8.2 Discussion

That biological erosion rate can be very rapid can be deduced from the data presented in *Table 4.4*. Not only is rock removed directly, but the rock is also weakened by boring and is therefore more readily removed by wave action, especially during storm events.

TABLE 4.4 BIOLOGICAL EROSION RATES (MOSTLY TROPICAL SPECIES) (FROM KIRK, 1975; TRUDGILL 1976b)

Organism	Rock	Rate ($mm\ a^{-1}$)
Borers		
Clam (*Penitella*)	Mudstone	12.0
Sponge (*Cliona* sp.)	Limestone	14.0
date mussel (*Lithophaga*)	Limestone	8.0–9.0
Limpet (*Patella*)	Limestone	0.25–1.0 (home-scar)
Grazers		
Winkle (*Littorina* sp.)	Limestone	0.6
Chiton (*Acanthopleura* sp.)	Limestone	0.5–1.0

4.9 ABRASION EXPERIMENTS

Abrasion can be measured in both the fiela and laboratory. In the first case, rock tablets (described in section 3.6) can be made into two sets. The first is placed in a coarse mesh bag and the second is placed in finer meshes which exclude sand particles. Mass losses can then be compared when these are suspended on nylon rope in abrasive environments such as in streams where sand is present, or in coastal environments. The tablets in the smaller mesh will experience a small decrease in mass loss due to the restriction in water circulation because of the smaller mesh; however, differences in mass loss between the two bags greater than 10% can be ascribed to the effect of sand. Newson (1971, pp. 64–65) reports that for limestone systems studied in the Mendip Hills, Somerset, UK, abrasional mass loss was negligible at low flows, but at high flood flows when abrasive sand was entrained in the flow a 60% increase in mass loss occurred. It should be cautioned that tablets project into the flow and do not represent natural conditions, but they can provide a useful evaluative tool for assessing the relative importance of abrasion.

TABLE 4.5 MASS LOSSES OF ROCK CUBES DURING ABRASION IN ABRASION MILL FOR 48 h

Rock	Before abrasion (g)	After (g)
Quartzite	20.00	19.99
Coal measure sandstone	20.00	18.20
Carboniferous Limestone	20.00	14.30

Secondly, abrasion of different rocks can be assessed by using an abrasion mill (these are used for polishing semi-precious stones). The rocks can be cut into cubes and then their rounding measured using calipers and/or their mass loss can be monitored (though they will have to be dried prior to each measurement).

Examples of abrasion experiments are given in ***Table 4.5*** and ***Figure 4.14***.

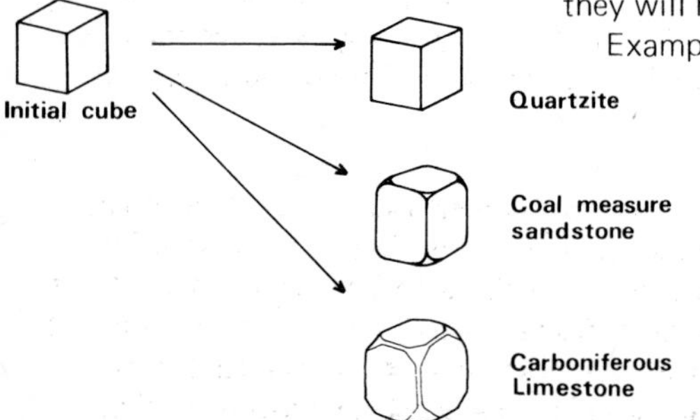

Figure 4.14 Abrasion of selected rock types

CHAPTER 5 SURFACE LOWERING OF UNCONSOLIDATED SEDIMENTS AND SOILS

5.1 INTRODUCTION

Soils, river-bank alluvium and soft cliff material can be eroded rapidly during intensive events such as storms and as such their erosion can often be readily measured. The erodibility of material can also be assessed with respect to a variety of erosion processes. Long-term trends can be measured using repeated surveys of a site, either by field work or by using data sources such as old maps or photographs.

5.2 RUNOFF PREDICTION

Surface erosion and transport is often effected by lateral surface-water movement. The occurrence of this flow may be predicted by a simple infiltration measurement. The method is designed to assess the mount of water that a soil can absorb under saturated conditions; any excess to this amount will tend to run overland and therefore encourage surface erosion. The excess water is termed 'infiltration excess overland flow'. In simple tests, water is added to the soil surface and the rate of infiltration is measured. The tests do not necessarily duplicate natural processes but they may be replicated at several sites for comparative purposes. The methods involve the ponding of water on the soil surface and monitoring the rate of water entry into the soil, firstly by simply assessing the rate of lowering of the water surface and secondly by monitoring the rate of water supply needed to maintain a constant ponded water level.

5.2.1 Method

A metal or plastic ring should be used, with a sharp lower edge that can be hammered into the soil. Tin cans are suitable, but may buckle easily during insertion. In addition, the ring should be as large as possible, preferably no smaller than 30 cm diameter. This is because there will be an edge effect as some rapid drainage may occur down the inside of the ring. A large diameter will minimise this effect by maintaining a high area:perimeter ratio. A robust PVC cylinder cut from piping, 30 cm diameter, 30 cm deep and of 0.5 cm

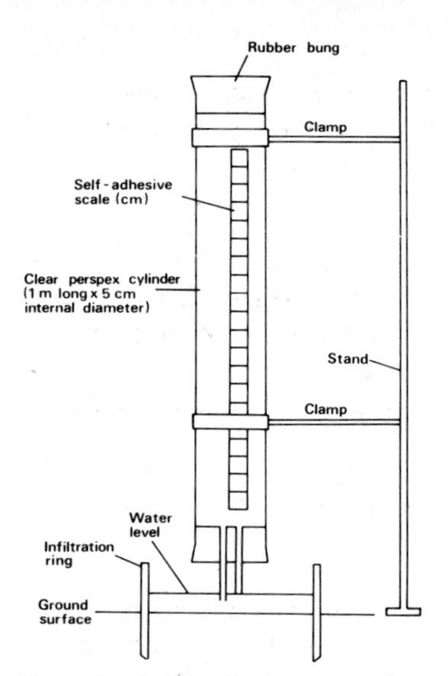

Figure 5.1 Infiltration apparatus used to maintain a constant water head (After Burt, 1978)

wall thickness, with a bevelled lower edge, provides an optimum ring. The inside of the ring should be graduated in mm (or a ruler may be held inside it) for the simple water-level fall test. The ring should be hammered well into the soil to ensure that leakage is minimised. The ring should then be filled with water from a bucket and the fall of water level noted after 10 s, 30 s and then 1 min intervals. However, as water level drops, the head of water decreases. Thus, the weight of water pressing down on the soil surface decreases and the infilitration rate will drop. This effect can be crudely controlled by topping the water level up to a specified level after each reading. Greater control is achieved by the use of a constant-head device (***Figure 5.1***). This is a graduated container for water inverted over the ring and with one long and one short outlet pipe. Air bubbles into the container through the shorter tube and water will run out from the longer. As the level in the ring drops, water will be replaced from the container. A constant water-pressure head is thus kept over the ponded soil surface and the infiltration rate is measured by monitoring the rate of fall in the bottle. The measurement of infiltration rate is taken as the steady-state value which is maintained for more than five minutes.

5.2.2 Discussion

The rate of water lowering, usually expressed in mm min^{-1}, predicts the maximum rate at which water may enter the soil under the conditions obtaining. It can be suggested that if rainfall intensities exceed infiltration capacity then overland flow is liable to occur — implying that surface-wash erosion can also occur. However, under natural conditions, infiltration capacity varies with soil moisture content and, in addition, the saturated, steady-state conditions used in the test do not simulate the conditions during the entry of natural rainfall into the soil. Nevertheless, the tests, in describing the maximum infiltration rate, set the upper limit for water entry into the soil and can be used for the identification of areas where overland flow is most liable to occur. Illustrative results are: ***pathway*** 0.01 mm min^{-1}, ***grassland*** 5 mm min^{-1}, ***freshly ploughed soil*** 20 mm min^{-1};

suggesting that in this case surface flow is very unlikely in the latter two cases (rainfall in the UK does not commonly exceed intensities of 5 mm min^{-1}) but that overland flow is very likely on the path, rainfall intensities of under 1 mm min^{-1} being common. The use of a recording rain gauge where rainfall intensities are recorded in millimetres on a time chart will be most useful in this kind of study.

It should be stressed that in soils of high infiltration capacity an adequate nearby water supply is essential, and to prevent the apparent infiltration rate merely being an artefact of the maximum delivery rate of the constant-head bottle tubes, the ring should firstly be well primed with buckets of water.

5.3 WASH TRAPS

The principle of wash traps is that they collect water which is running over the soil surface and any sediment which is being moved by that water.

5.3.1 Method

The trap is essentially a box sunk into the ground (*Figure 5.2*) where the junction between the soil and the trap should be a flush join. The sediment may be left to settle in a simple box trap or it may be led off to settling tanks. The amount of runoff water

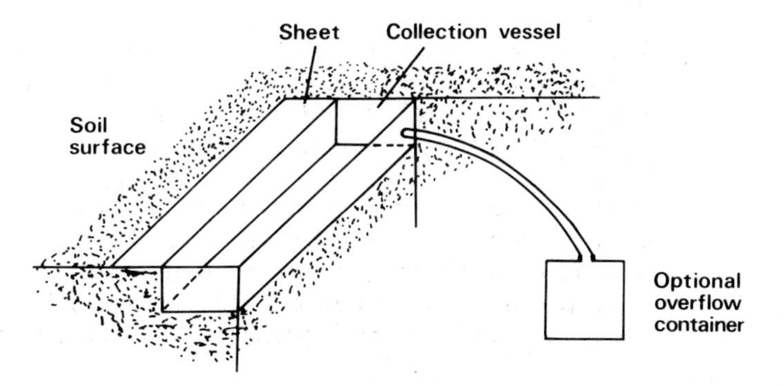

Figure 5.2 A wash trap for collecting sediment transported in overland flow

should be measured in litres and can be discarded after settling has taken place. Measurement of the amount of sediment is facilitated by having an inner removable container in the collection vessel. After discarding the clear water, the container and contents should be dried (preferably at 105°C till constant mass) and the container plus solid weighed. The dry mass of the eroded material can be calculated by subtracting the volume of the container (***see*** Briggs, 1977b, p. 72). The sediment yield can thus be expressed in terms of g ℓ^{-1} runoff.

Sediment yield can be expressed in terms of g m^{-2} if the source area for the trap is bounded by boards or other edging material capable of cutting off the flow from adjacent areas. A bounded plot 1 m wide at the base may be constructed. This may be bounded laterally, for example for 10 m up slope, and also at the top.

5.3.2 Discussion

Experimental design can be arranged to compare sediment yield from vegetated and unvegetated plots on similar soils on similar slopes; alternatively, position on slope could be varied, holding vegetation and soil type constant.

Settling times of the order of 2–3 days may be needed if clay-sized particles are present in the runoff. Light organic matter particles will not settle and can be strained from the water by sieving and measured separately if desired.

Some data for wash-trap collections are shown in ***Table 5.1*** for a loessial sandy soil in West Belgium (after Bollinne, 1976). The experimental design was arranged to test the hypothesis that bare soil would lead to greater erosion losses than those soils with a vegetated cover. In fact, the data show a great variability between the years of observation (in relation to variations in rainfall) but in the years when substantial losses

TABLE 5.1 WASH-TRAP SEDIMENT FROM THREE PLOTS ON LOESSIAL SANDY SOIL (TONNES ha^{-1}), MARCH-OCTOBER PERIOD

	Sugarbeet	Winter wheat	Fallow
1973	0.4	0.3	0.0
1974	30.1	4.4	82.2
1975	1.0	0.7	6.1

TABLE 5.2 OBSERVED RATES OF SURFACE WASH (AFTER YOUNG, 1974)

Climate	Vegetation	Slope angle (degrees)	Volumetric movement ($cm^3\ cm^{-1}\ a^{-1}$)	Ground lowering (B*)	Location
Polar	Bare	–	–	1–10	Spitzbergen
Temperate maritime	Grass	25	0.08	–	Northern England
Temperate maritime	Grass	–	0.09	–	Scotland
Temperate maritime	Tussock grass	–	–	10	New Zealand
Temperate maritime	Tussock grass	22–32	0.13	2	New Zealand
Temperate maritime	Tussock grass	36	5.6	19	New Zealand
Temperate maritime	Bare	5	–	10 000	Wales
Temperate maritime	Bare	–	1.5	–	Wales
Temperate continental	Bare	43	–	>230 000	New Jersey, USA
Temperate continental	Grass	–	–	5–0.05	Poland
Temperate continental	Grass	–	–	5	Poland
Temperate continental	Mainly grass	–	–	10–60	USA
Temperate continental	Forest	–	–	0.03	Poland
Temperate continental	Bare	36–39	–	100	Poland
Temperate continental	Forest	–	–	0.1	Poland
Temperate continental	Forest, grass	21–23	–	0.508	New Mexico, Wyoming, USA
Temperate continental	Bare	steep	–	900	Alberta, Canada
Temperate continental	Bare	45–50	–	100 000	Crimea, USSR
Warm temperate	–	–	–	50–100	N.S.W., Australia
Humid sub-tropical	–	–	–	19	Mississippi, USA
Mediterranean	–	–	–	90	Southern France
Mediterranean	–	29–35	–	29	California, USA
Mediterranean	–	42	–	253	California, USA
Mediterranean	Garrigue	14	–	0.4	Southern France
Mediterranean	Garrigue	25–35	–	75	Southern France
Semi-arid	–	–	–	2 000	Colorado, USA
Semi-arid	–	steep	–	6 400–8 200	New Mexico, USA
Semi-arid	Bare	–	4.2	–	Arizona, USA
Semi-arid	Pinon, sage	23–35	–	7 600–11 700	New Mexico, Wyoming, USA
Arid	Bare	–	0.3	–	Arizona, USA

continued overleaf

* B = Bubnoff Unit = 1 $\mu m\ a^{-1}$ = 1 $m\ ka^{-1}$ = 1 $m^3\ km^{-2}\ a^{-1}$

TABLE 5.2 continued

Climate	Vegetation	Slope angle (degrees)	Volumetric movement ($cm^3\ cm^{-1}\ a^{-1}$)	Ground lowering (B*)	Location
Savanna	Savanna	–	–	1.6	Senegal
Savanna	Savanna	–	–	39	N.T. Australia
Savanna	Cerrado, grass	–	0.6–9.6	–	Mato Grosso, Brazil
Rainforest	Forest	–	–	500–1 500	Ivory Coast
Rainforest	Forest	15–22	1	–	Malaya
Rainforest	Forest	26–30	4	–	Malaya
Rainforest	Montane forest	steep	–	260	Tanzania

* B = Bubnoff Unit = 1 $\mu m\ a^{-1}$ = 1 $m\ ka^{-1}$ = 1 $m^3\ km^{-2}\ a^{-1}$

did occur, losses from bare soil and a widely spaced crop (sugar beet) were greater than those under a cover of winter wheat.

The role of vegetation in influencing the rates of surface wash was also emphasised by Young (1974) who compiled the data shown in ***Table 5.2***. For comparable environments, the rates for bare areas are clearly greater than those for the vegetated areas.

5.4 SPLASH BOARDS AND SPLASH TRAPS

As well as movement in surface wash (5.3) particles may be detached from the surface and moved by the impact of raindrops. The principle of the splash board is that material thrown upwards and sideways by rain splash is intercepted by a vertical board.

5.4.1 Method

The technique can be used to illustrate the effects of splash or to measure the amounts moved. If the former is required, a board 1 m high X 0.5 m wide is taken and large sheets of blotting paper, or other white paper, should be affixed to the board so that any splashed material will be visible. The board is then mounted vertically in the soil, emplacing the lower 10–20 cm of the board in the soil so as to achieve stability.

Several boards can be left out in position until rainfall occurs. Alternatively, water can be applied artificially in standard form: crudely, using a watering can with a standard amount of water applied from a standard height, or, with greater control of water droplet size, by the use of a sprinkler device (***see*** section 5.5). In either case, effective demonstrations can be made of the differences between splash under different types of cover, for example between grassland, ploughed land and a forest floor on similar soils. The

(*a*) Board with blotting paper attached

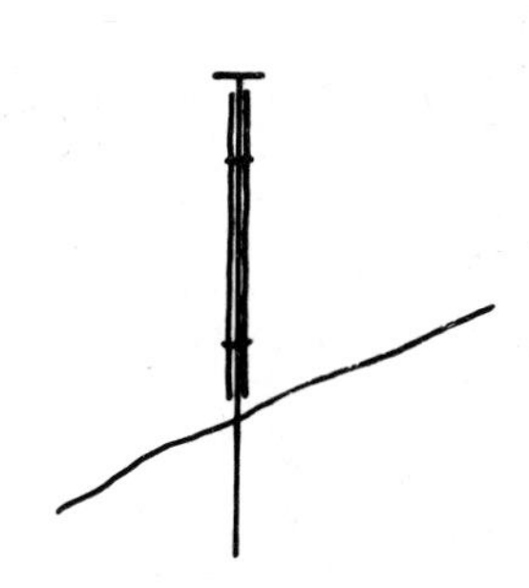

(*b*) Simple board with trough

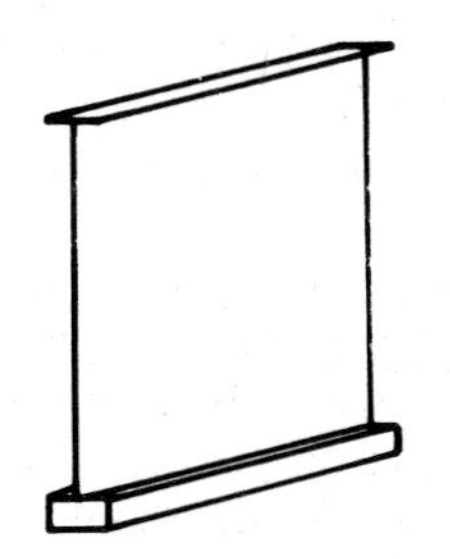

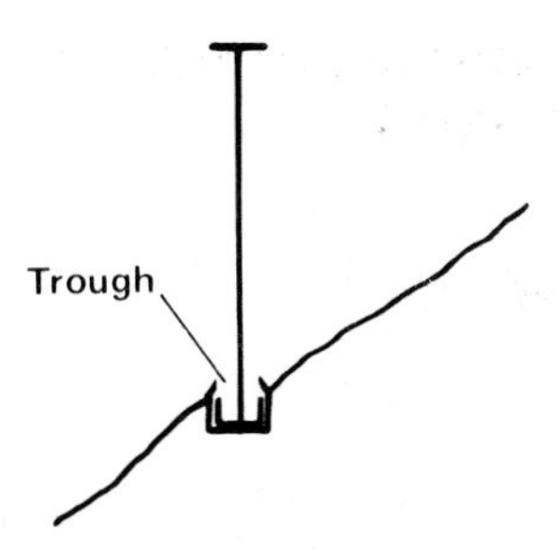

Figure 5.3 A splash board

height of splash, the size of splashes and grain size may be seen by inspection (it is better to use blotting paper for this purpose as it minimises runoff down the board).

Quantitative assessment of the amounts of soil moved involves the collection of the splashed material. This can be achieved by the use of a trough at the base of the board into which the material will run (***Figure 5.3***). The board should also be rinsed down after use and the trough should have a small lip to minimise splash out from the trough. The material collected in the trough may be treated in the same manner as that for wash traps (***see*** section 5.3) and, again, removal and treatment is facilitated by the use of an inner removable container in the trough.

5.4.2 Discussion

In terms of geomorphological processes, a major significance of the technique is the assessment of the amount of soil material which can be splashed downslope, thereby facilitating slope erosion. Splash boards can be placed across the slope, so that the face of the board is at right angles to the direction of the slope, in order to compare the amounts

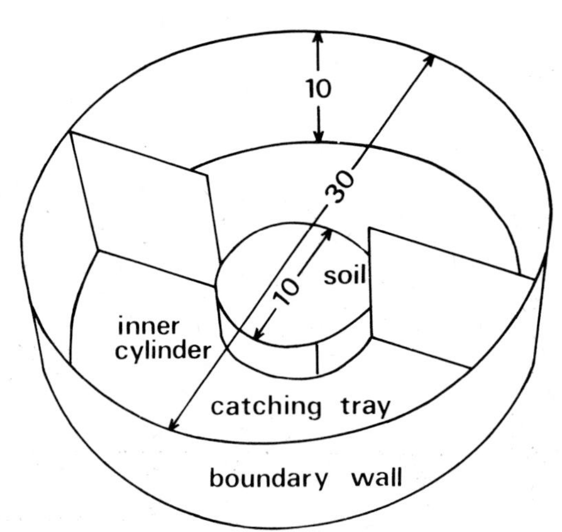

Figure 5.4 Small-cup splash erosion trap, dimensions in cm (After Morgan, 1978)

TABLE 5.3 RATES OF SPLASH EROSION (FROM MORGAN, 1978)

Site	Slope angle (degrees)	900–Day total	Total (kg m^2 a^{-1})
Top-slope	9	8.24	2.73
Mid-slope	11	20.53	6.81
Lower slope	11	15.41	5.11

of downslope and upslope movement of splashed material. Using a simple splash cup (*Figure 5.4*), Morgan (1978) showed how downslope movement of splashed material was greater with increase of slope angle from 9 to 11 degrees (*Table 5.3*). A similar relationship is shown by Kwaad (1977) (*Figure 5.5*). The use of sprinklers in conjunction with splash boards is discussed below (section 5.5).

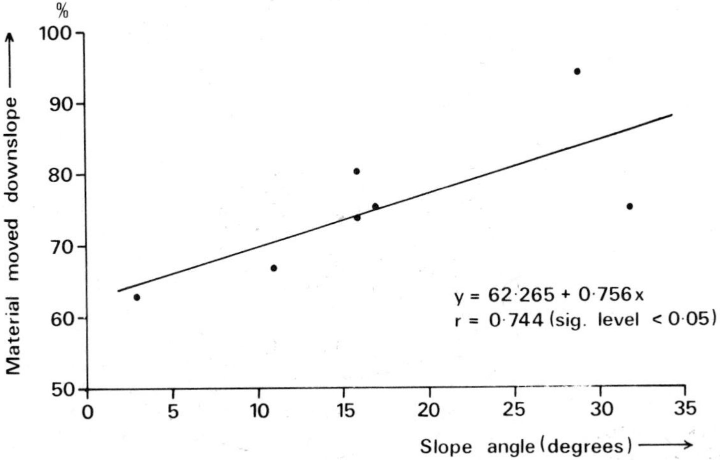

$y = 62{\cdot}265 + 0{\cdot}756x$

$r = 0{\cdot}744$ (sig. level $< 0{\cdot}05$)

Figure 5.5 The relation between splash erosion and slope angle (After Kwaad, 1977); amount of material moved downslope as a percentage of downslope and upslope total

5.5 SPRINKLERS

Sprinklers can be used to simulate rainfall for use with splash boards (section 5.4) and also wash traps (section 5.3).

5.5.1 Method

Sophisticated devices can be made to simulate the drop sizes and impacts of rainfall as closely as possible (e.g. Tricker, 1979; Riezebos and Seyhan, 1977) but relatively simple alternatives can also be made.

Drop sizes from domestic watering cans are usually much larger than natural raindrop sizes and thus do not simulate the kinetic energy of rainfall, but at a simple level of analysis, they may be used to compare one soil type with another under standard, if artificial, conditions. Alternatively, finer droplets and lower delivery rates may be produced by the use of hand-pumped garden sprays. Comparable results may be obtained if the same spray setting is used for every experiment. Sprinkler attachments to garden hoses can also be useful for larger areas.

A relatively simple device, but one which produces drop sizes approximately the same as natural rainfall, consists of a network of capillary tubes connected to a reservoir container. The water passes through a bifurcatory branching distributary network and drops from fine capillaries or hollow needles. Such a device is illustrated in ***Figure 5.6***, with a network of 1 mm interval diameter tubes, and with the drops dispersed by a 3 mm mesh screen. In the field the reservoir can be supported on a portable framework such as a step ladder. Water can be dribbled onto trays of earth in the laboratory, using either intact blocks or loose clods from an arable field, and dribbled onto ***in situ*** soil in the field. The amount of splash erosion can be assessed using splash boards set up to one side of the splash zone or in the centre of the zone.

5.5.2 Discussion

Examples of results from the use of a portable sprinkler and splash boards are shown in ***Table 5.4***. Firstly, the influence of a natural leaf litter cover is tested. The mean value for a litter-covered area is 0.0211 g and for a bare area, 0.2747 g. Secondly, the effect of slope angle is tested. Markedly greater net downslope splash occurs at higher slope angles: splash erosion increases from 0.009 g to 0.0799 g under a litter cover when slope angle increases from 1 to 38 degrees and from 0.0003 g to 1.0333 g on bare soil when the slope increases from 1 to 40 degrees.

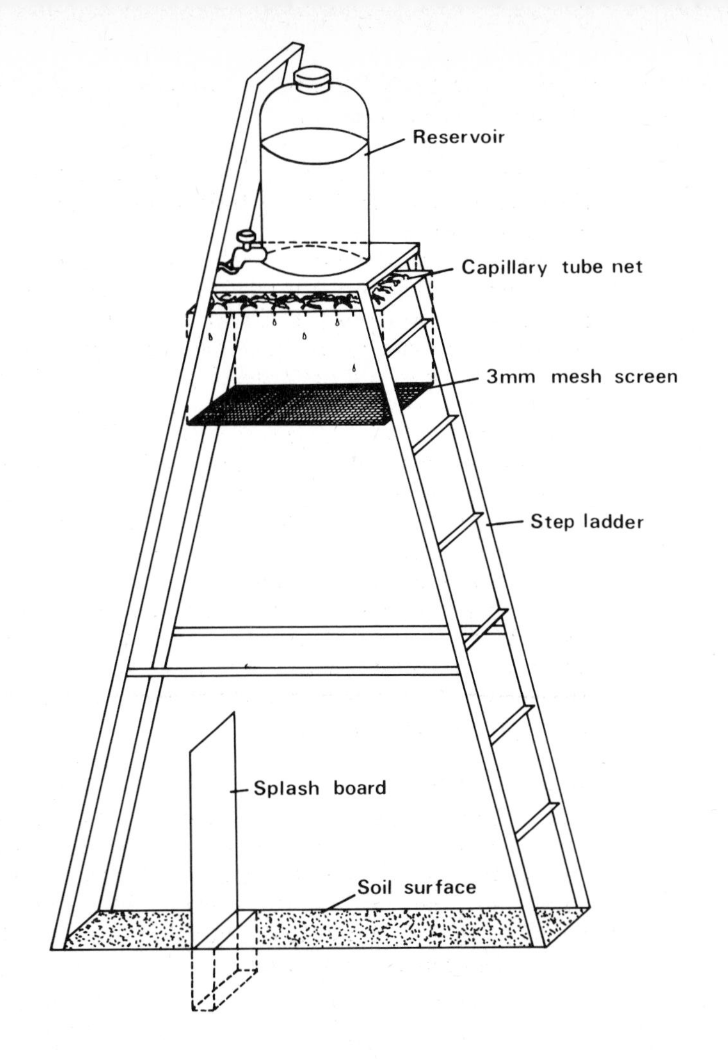

Figure 5.6 Sprinkler device with splash board (After Van Zon, 1979)

TABLE 5.4 RESULTS FOR SPLASH EROSION USING 'STEP LADDER' RAINFALL SIMULATOR (AFTER VAN ZON, 1979)

Condition of surface		Slope angle (degrees)	Mass of material splashed (g) Downslope	Upslope	Net downslope
1	Natural*	1	0.0180	0.0189	0.0009
2	Natural*	11	0.0119	0.0101	0.0018
3	Natural*	20	0.0144	0.0067	0.0077
4	Natural*	29	0.0199	0.0048	0.0151
5	Natural*	38	0.0936	0.0137	0.0799
6	Bare	1	0.1303	0.1300	0.0003
7	Bare	10	0.0408	0.0266	0.0142
8	Bare	20	0.0199	0.0178	0.0021
9	Bare	30	0.3538	0.0301	0.3237
10	Bare	40	1.1835	0.1502	1.0333

*litter intact

5.6 EROSION PINS

Measurements of infiltration capacity (5.2) can assist with the prediction of runoff, and the use of wash traps (5.3), splash boards (5.4) and sprinklers (5.6) can be used to assess sediment-transport processes. In conjunction with these measurements, there is a need to measure the rate of lowering of the actual surface of the soil. The use of an erosion pin is based upon the employment of the head of a large steel or wooden pin as a reference point from which lowering of material around the pin can be measured. In addition to the lowering of soil surfaces, the retreat of vertical surfaces, such as river banks, can also be measured.

5.6.1 Method

A rod made from steel, wood or rigid plastic is hammered into the material to be measured, usually with the head flush to the surface; alternatively the head may be left standing up from the surface for ease of relocation and the initial height difference between the pin head and the surface measured. Successive lowering of the surface relative to the head can be monitored.

Dimensions of 30–50 cm length or more are needed and diameters of 0.2 to 0.5 or

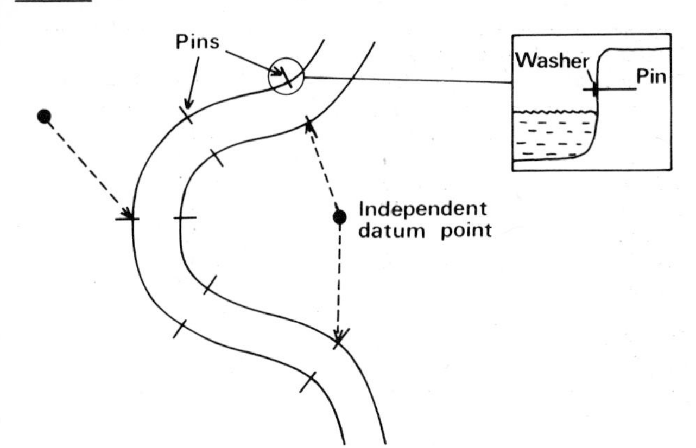

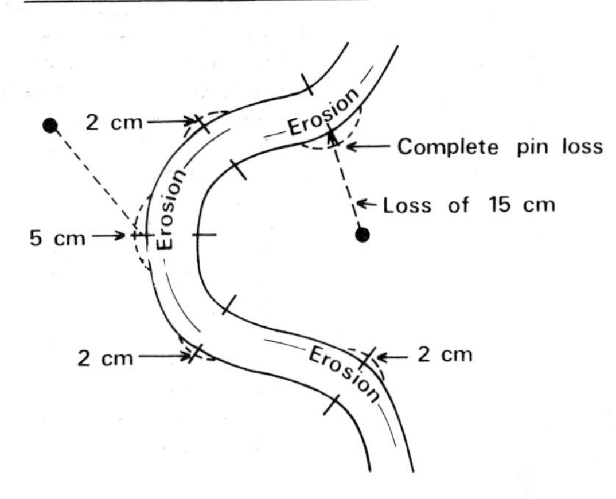

Figure 5.7 Bank erosion survey, using pins

1 cm are usual. The pin should be as thin as possible as the presence of a pin causes an obstruction and the resulting turbulence could increase the rate of erosion. Pins should also be as strong as possible to withstand water flow. Care should be taken during insertion not to loosen or compact the soil. Some workers have used washers on the pin in order to facilitate a decision as to where the surface lies, but in a river-bank situation, these can lead to increased turbulence.

The locations of erosion pins should be surveyed to a nearby datum point such as a tree or a building, especially in the case of stream banks where bank collapse may lead to the loss of the pin (*Figure 5.7*). If a datum point has been established, the extent of the collapse can still be measured despite the loss of the pin. Care should be taken during installation and reading to avoid drowning accidents; work should not be attempted at high flows when bank collapse is likely.

5.6.2 Discussion

Surface lowering pins are described in detail by Haigh (1977) and the method used is shown in *Figure 5.8*. Haigh also provides a survey of published erosion-pin data which are

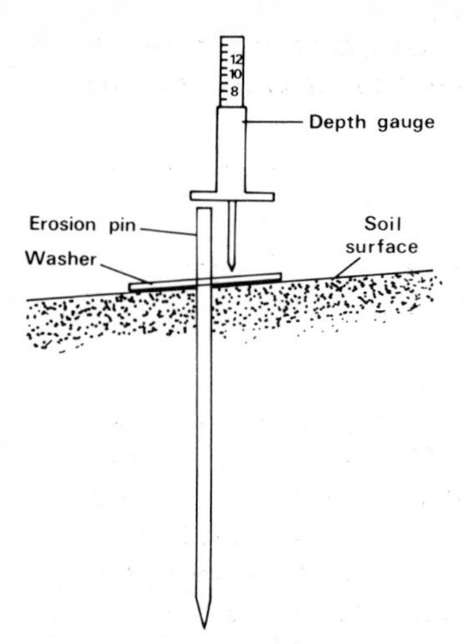

Figure 5.8 Erosion-pin device (After Haigh, 1977)

TABLE 5.5 EROSION-PIN DATA (FROM HAIGH, 1977)

Location	Results
Nebraska, USA, Sioux Co. (Badlands)	8–38 mm a^{-1} (slopes) 15–46 mm a^{-1} (divides)
New Mexico, USA (Upper Arroyo de los Frijoles)	23 mm a^{-1} (61 pins) 44 mm a^{-1} (57 pins) 2 mm a^{-1} (19 pins)
Wyoming, USA, Hudson, Last Day Gully	3.3 mm a^{-1} standard deviation 8.2 mm a^{-1}
North Dakota, USA, Little Missouri (Badlands)	9 mm a^{-1} on S.W.-facing slopes 4 mm a^{-1} on N.E.—facing slopes
South Wales, Lower Swansea Valley	Average: 10.6 mm a^{-1} 5 degree slopes: 10 mm a^{-1} 45 degree slopes: 32 mm a^{-1} 62 degree slopes: 43 mm a^{-1}
South Wales, Blaenavon (colliery spoil mounds and infilled opencast sites)	Vegetated slopes: 3.1 mm a^{-1} (Unvegetated slopes: 4.6 mm a^{-1})
North Yorks Moors, UK, Hodge Beck	Bare mineral soil: 63.9 mm a^{-1} Bare peat: 40.9 mm a^{-1}
Georgia, USA, Catersville (artificially devegetated highway cuts)	23–53 mm a^{-1} on a 19.5 degree slope. N.W.—facing plots lost more than twice the amount lost by plots facing S.E.

summarised in ***Table 5.5***. The effects of a lack of vegetation and steep slope in increasing erosion rate predicted in sections 5.3–5.6 are readily seen from these data. Data on river-bank erosion are also given in section 5.8 (***Table 5.8***).

5.7 PROFILE SURVEY

Profile survey involves the repeated measurement of surface levels and cliff edges in order to monitor the rates of retreat, and also deposition, during the redistribution of unconsolidated material.

5.7.1 Method

Profile survey can be carried out at a number of different scales, either at the small scale, measuring profile changes of a few millimetres or centimetres (***Figure 5.9***) or at a larger scale using standard surveying techniques such as levelling. Abney levels may be employed at a simple level of investigation and quickset levels provide a more accurate method. In the case of the former a row of poles of equal height is placed along a profile, especially at changes of slope, and the distances between the poles are measured. Abney levels are placed at the top of the poles (or at a standard height on the pole) and the angle to the top of the next pole (or standard height) is measured. The profile can then be drawn out on graph paper at an appropriate scale using a protractor. If poles are not available, sightings may be taken from one person to another. The height of the eye level of the first person reading the Abney should be established on the second person so that the line of sight is parallel to the ground surface.

Great care should be taken when surveying steep and unstable profiles, especially on cliffs, where safety helmets should be worn. Beware of the possibilities of sudden falls, mudflows, and being cut off by the incoming tide in coastal cliff areas.

5.7.2 Discussion

Surveying is useful for monitoring coastal processes such as changes in beach profiles in relation to storm events (Buchan & Ritchie, 1979) (***Figure 5.10a***) and changes in cliff profiles in response to winter storms (***Figure 5.10b***). Although partly concerned with examples of hard rock (***see*** chapter 4) Kirk (1975) gives a range of data for cliff retreat from profile survey and also old maps and photographs (***see*** section 5.8), and these data

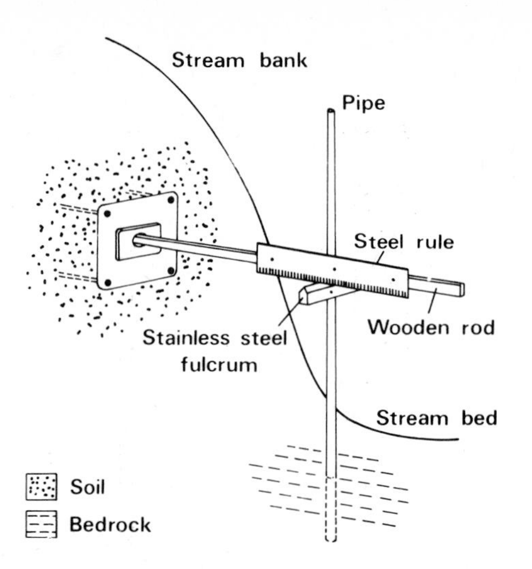

Figure 5.9 Bank profile survey device (After Finlayson, 1977)

are shown in ***Table 5.6.*** While the sites vary in exposure, it is clear that unconsolidated material such as glacial drift, siltstones and slumped mudstones and sandstones can be eroded rapidly at rates ranging between approximately 1.0 and 2.0 m a^{-1}. Experimental design for projects could compare (1) profile retreat on different deposits under similar conditions of exposure or (2) the effects of different exposures on similar substrates; in either case enabling the testing of hypotheses concerning the roles of the exposure and substrata factors involved.

Much erosion is achieved by the occurrence of extreme events but it may be the case that, over a long time span, frequent small erosion events may achieve more erosion than infrequent events of a large magnitude. ***Table 5.7*** demonstrates the relationship between magnitude and frequency of events measured by Hooke (1980) for river-bank erosion on the river Exe. The data are for river-bank erosion assessed from repeated surveys of meander position from datum points and also from the use of old maps (section 5.8). Small amounts of erosion are achieved by low flows (e.g. 4.4 cm retreat at up to 50 m^3

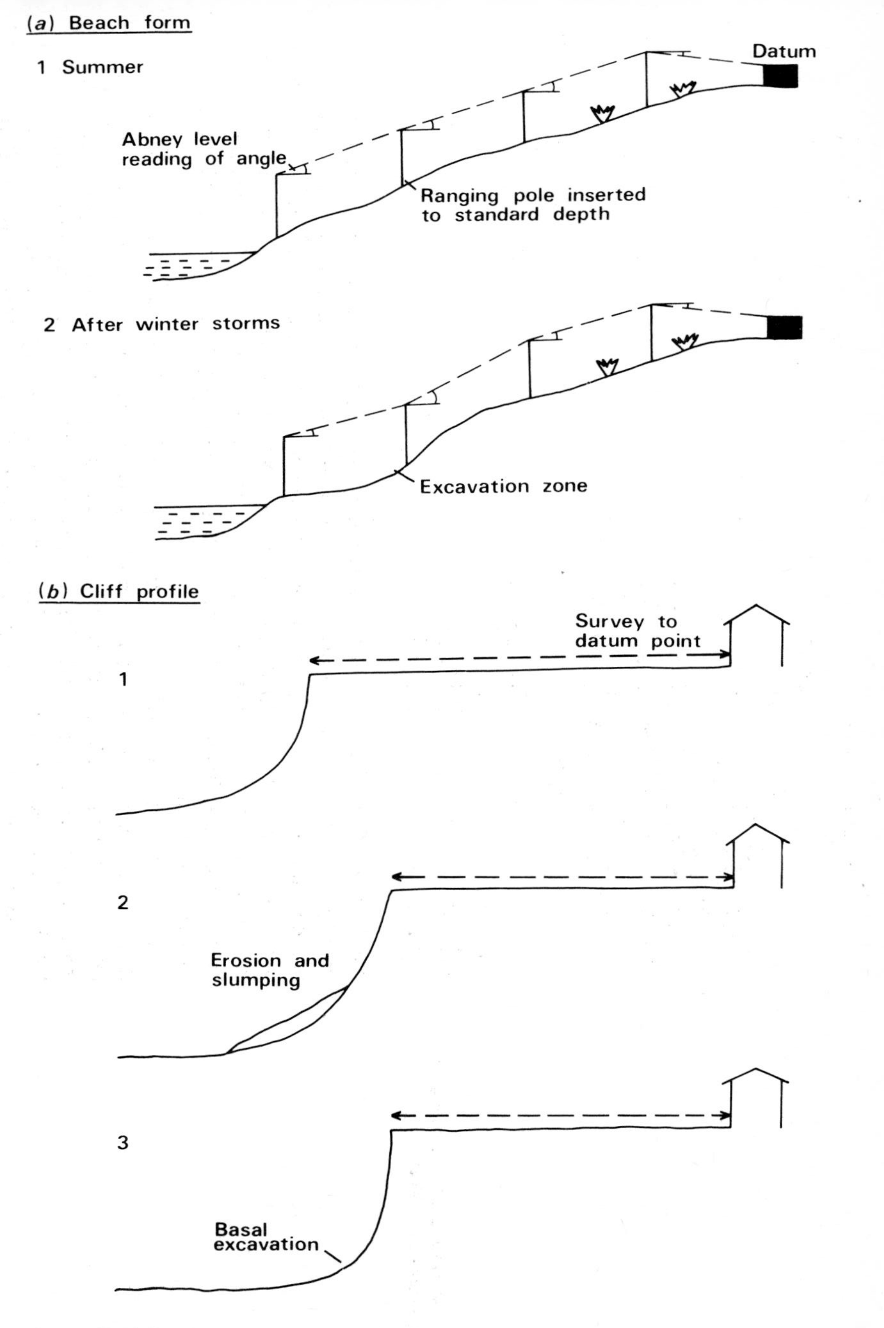

Figure 5.10 Profile survey

TABLE 5.6 RATES OF CLIFF RETREAT (FROM KIRK, 1977)

Method	Rate ($m\ a^{-1}$)	Erosion process	Lithology	Location
Surveys	zero	—	Basalt	Victoria, Australia
Surveys	0.04	Slumping and cave collapse	Aeolianite	Victoria, Australia
Surveys	0.008	Weathering and abrasion	Arkose	Victoria, Australia
Surveys	0.0175	Weathering and abrasion	Siltstone	Victoria, Australia
Old photographs	0.0127	Abrasion	Chalk	Vale of Glamorgan, Wales
Old photographs, volume calculation	0.004–0.006	Abrasion and mass movement	Marls and Limestones	Gotland, Baltic Sea
Comparison of maps	0.202	Undercutting and slumping	Glacial drift	W. Washington, USA
Air photographs	0.1–1.01	Wave action and mass movement	Sandstone	Japan
Surveys and air photographs	0.1–5.05	Wave action and mass movement	Range from sands to sandstone	Japan and worldwide
Surveys and photographs	0.89	Wave action and mass movement	Siltstones and sandy mudstones	Taranaki, New Zealand
Surveys and photographs	0.69	Wave action and mass movement	Siltstones and sandy mudstones	Taranaki, New Zealand
Surveys	1.69	Wave action on mass movement deposits	Slumped mudstones and sandstones	Gisborne, New Zealand
Repeated surveys	1.86–2.0	Wave action	Glacial drift and sands	E. Coast, Lake Michigan

TABLE 5.7 DATA FOR RIVER BANK PROFILE SURVEY, RIVER EXE, UK (FROM HOOKE, 1980), EROSION IN RELATION TO RIVER DISCHARGE AND FREQUENCY OF OCCURRENCE (R.I. = RECURRENCE INTERVAL OF FLOW LEVELS)

Flow ($m^3\ s^{-1}$)	R.I. (a)	Frequency (days a^{-1})	Erosion (cm)	% of total eroded
50	0.0786	12.7200	4.4	47.5
80	0.1432	6.9800	14.5	85.7
100	0.2135	4.6800	21.2	84.1
120	0.3183	3.1400	27.9	74.2
150	0.5794	1.7300	37.9	55.6
200	1.5700	0.6370	54.7	29.5
250	4.2700	0.2342	71.4	14.2
350	31.4300	0.0318	104.9	2.8
450	231.5000	0.0043	138.4	0.5

s^{-1} flow) but because they are more frequent they achieve 47.5% of the overall erosion. Large flow events achieve massive bank erosion, but with a recurrence interval of around 200 years, the proportion of overall erosion achieved is small.

5.8 PHOTOGRAPHS AND MAPS

Observations using profile surveys are limited to the time span available to the observer. Many major geomorphological erosion events occur only infrequently, however, and it is easy to miss the effects of rare events. Therefore reference may usefully be made to old photographs and maps in order to plot major changes in the recent past. The availability of old photographs and maps will obviously vary locally, but archives can often be consulted in local libraries and universities. Frequently, old newspaper reports of cliff falls can be used to illustrate the frequency of events. Photographs may show the relationship of cliff faces to objects which still exist, and if the photograph is dated then the amount of erosion which has occurred since the photograph was taken can be estimated. Similarly, old maps can be very useful in revealing former cliff lines and former courses of rivers in relation to fixed points which can still be identified.

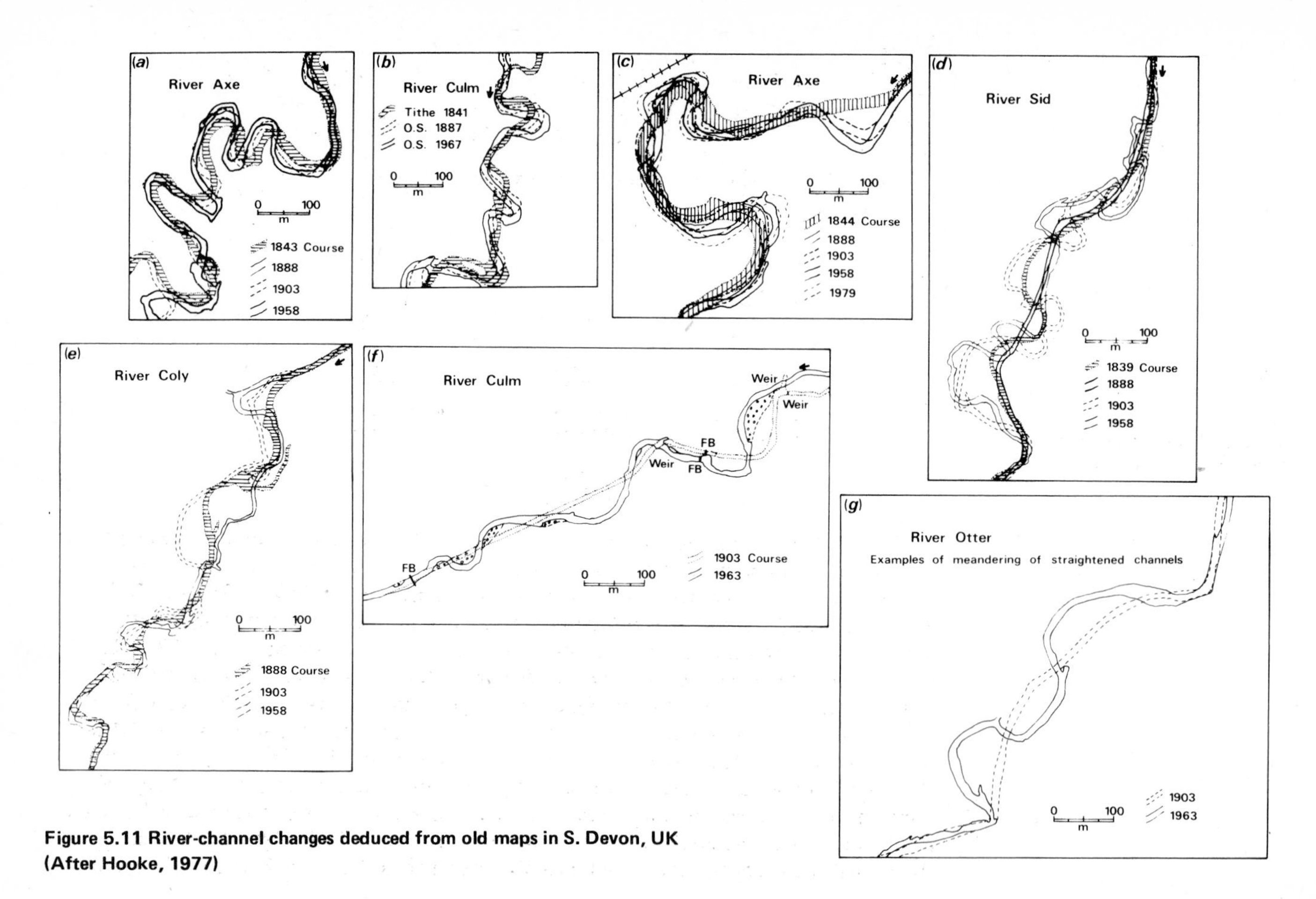

Figure 5.11 River-channel changes deduced from old maps in S. Devon, UK (After Hooke, 1977)

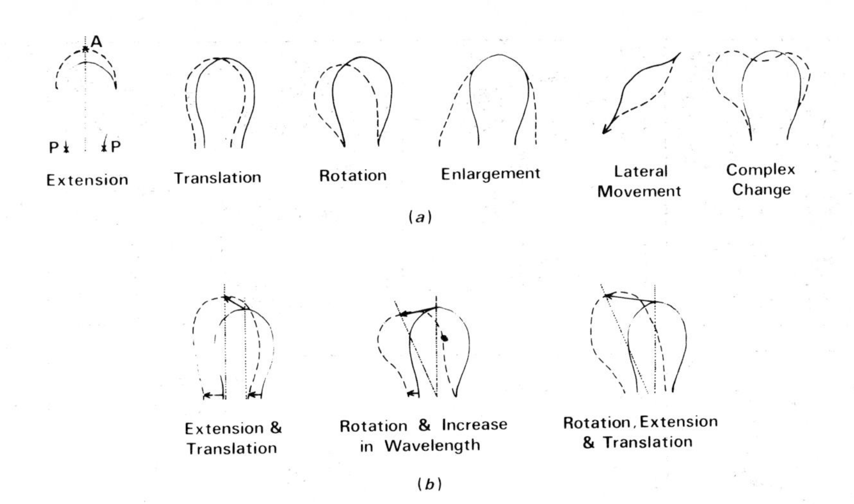

Figure 5.12 Types of river-channel changes (After Hooke, 1977)

Coastal changes have been documented by Gibb (1978) and Kirk (1975). Gibb used maps and air photographs to compare erosion rates of 2.25 m a^{-1} on a mudstone with 3.46 on a conglomerate. Kirk used two sets of air photographs taken in 1942 and 1974. He identified cliff losses of up to 48 m in this time using this method, with up to 40 m accretion at other sites. Similarly, Hooke (1977) used old maps dated from 1839 to 1979 to plot changes in river courses in Devon. ***Figure 5.11a–g*** shows the variety of changes which she was able to plot and the types of changes identified are shown in ***Figure 5.12***.

Collations of data on river-bank erosion from a variety of sources are shown in ***Table 5.8*** (from Hooke, 1980), using maps, field measurements, erosion pins and photographs. Bank erosion rates are likely to increase with increasing discharge. In turn, discharge will increase with increasing drainage-basin area. Therefore, Hooke tested the relation between bank erosion and catchment area by plotting the data shown in ***Figure 5.13.*** This clearly shows that the larger the catchment area, the greater the rates of bank erosion. This is a

TABLE 5.8 PUBLISHED RATES OF BANK EROSION (FROM HOOKE, 1980)

River and location	Catchment area (km^2)	Mean discharge ($m^3\ s^{-1}$)	Rate of movement ($m\ a^{-1}$)	Period of measurement	Method
Ohio R, Kentucky, USA	–	–	0.357	1807–1958	Maps
R. Endrick, Scotland	97.66	6.94	0.5	1896–1957	Maps, mean
White R., Indiana, USA	6 042	66.2	0.67	1937–1968	Maps, mean
R. Mississippi, USA	–	–	23	1722–1971	Maps and historical data
R. Brahmaputra, India	934 990	1898	6–275	1952–1963	Maps
			15–792	1944–1952	Maps
R. Pembina, Alberta, Canada	–	19.2	3.35	1910–1956	Maps
			0.3		Bedrock channel
Little Missouri R., Dakota, USA	–	16	1.7–7.0	100+ years	Cottonwood trees
Des Moines R., Iowa, USA	–	–	6.6	1880–1970	Maps
R. Beatton, British Columbia, Canada	16 000	225	0.48	250 years	Trees and deposits
Crawfordsburn R., N. Ireland	3	–	0–0.5	1966–1968	Erosion pins
Clady R., N. Ireland	4	–	0–0.064	1966–1968	Erosion pins
R. Cound, Shropshire, England	100	4	0.64	1972–1974	Pegs, mean
R. Mississippi, USA	–	–	14.9–40.5	1963–1970	Field measurement
Wisloka R., Poland	–	22.5	8–11	1970–1972	Field measurement
Dunajec R., Poland	–	40	0.4–1.0		Maps
R. Bollin-Dean, Cheshire, England	120	–	0.01–0.09	1967–1969	Erosion pins
R. Ob, USSR	–	1434	0–15	1897–1958	Maps
R. Hernad, Czechoslovakia	5 400	10–30	5–10	1937–1972	Maps, mean
R. Rheidol, Wales	179	–	1.75	1951–1971	Maps, maximum
R. Tyfi, Wales	633	–	2.65	1905–1971	Maps
Chemung R., Pennsylvania, USA	–	–	3.05	1938–1955	Air photos
R. Klaralven, Sweden	5 420–11 820	650	1.6	1950–1956	Resurvey
			0.23	1800–1850	Maps
R. Torrens, S. Australia	78	–	0.58	1960–1963	Erosion pins

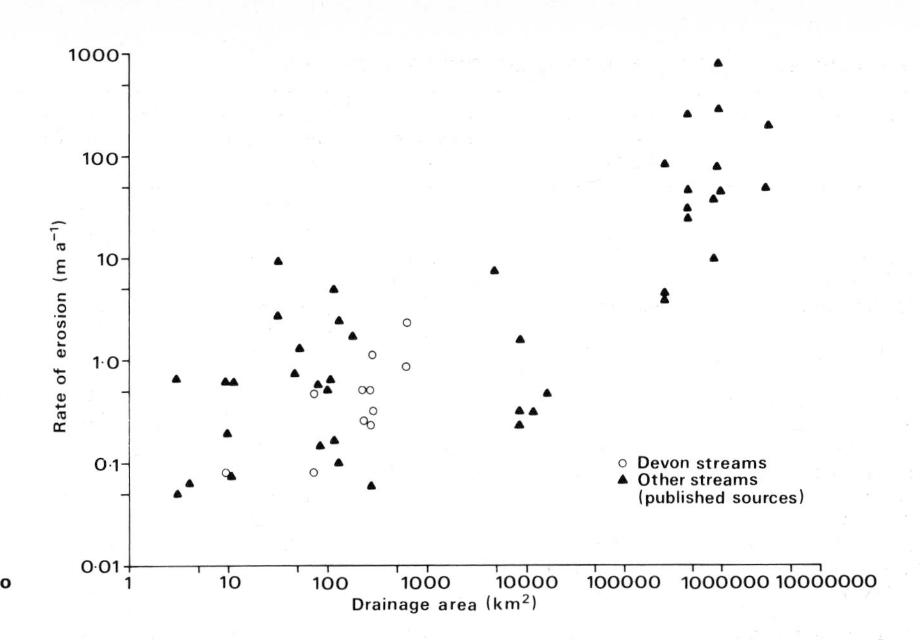

Figure 5.13 Relation of bank erosion to catchment area (After Hooke, 1980)

general relationship which could be applied to other sites by measurements of catchment area in order to predict rates of bank erosion.

5.9 TREE ROOT OBSERVATION AND TREE RINGS

The nature of tree growth may be used as indications of the stability of the landscape and, if the tree can be aged, then the timing of stability or instability may be estimated. For example, a scree is an unstable landform, but the presence of vegetation on a scree usually indicates that the landform has become less active than formerly. Conversely, vegetated areas may become less stable and progressive exposure of tree roots may be assessed. Work by Carrara and Carroll (1979) showed how tree roots may become exposed by soil loss during surface wash. Not only can simple observations be made of soil loss relative to the age of the tree, but the reaction of the tree roots may also be studied to give a more precise indication of when the erosion occurred. ***Figure 5.14*** shows how a root initially covered in soil has grown only on the lower side after exposure due to soil erosion. This growth is termed reaction wood and the exposure date can be calculated from the age of the reaction wood by counting the number of annual rings in the reaction wood. ***Table 5.9*** shows a summary of erosion-rate data gained by such an

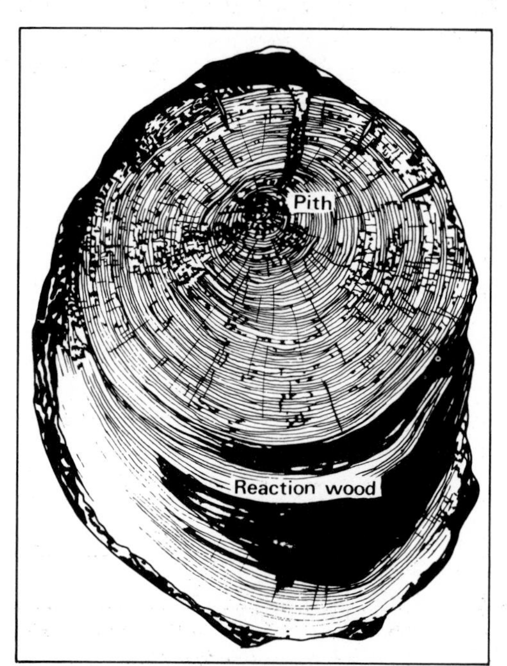

Figure 5.14 Root-growth distribution due to soil erosion (After Carrara and Carroll, 1979)

TABLE 5.9 EROSION RATES FOR THE LAST 400 YEARS, FROM ROOT-EXPOSURE DATA (AFTER CARRARA AND CARROLL, 1979)

Period (years ago)	0–99	100–199	200–299	300–399
Erosion rate (mm a^{-1})	1.79	0.49	0.33	0.22

analysis of trees of different ages and root exposure. The data demonstrate the increase in erosion in the last 100 years, using the Pinyon Pine (***Pinus edulis***) in Colorado, USA. ***Figure 5.15*** shows the complete data set and the trend of increased recent erosion rates is clearly seen. This type of approach can be used to assess accelerated erosion and increased landscape instability due to such processes as overgrazing.

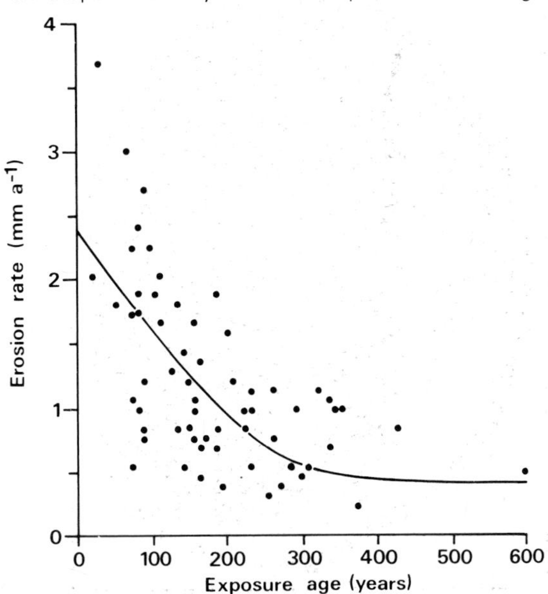

Figure 5.15 Erosion rates deduced from progressive root exposure (After Carrara and Carroll, 1979)

Tree ageing can also be used to estimate the minimum age of stability of screes. Tree age can be measured by coring the tree with a tree auger and counting the annual rings, or in the case of younger pine trees, by counting the number of annual growth nodes. The inference is that tree establishment could not take place on active screes. A minimum age is indicated, since a scree may be stable for some time before tree establishment occurred. It should also be stressed that in some situations screes may be built up round existing mature trees and careful observation will be needed to deduce whether this may be the case.

Such observations, together with general observations on overall vegetation cover, are of great assistance in deciding whether processes active today are leading to the current formation of landforms or whether the landforms observed are essentially fossil ones.

CHAPTER 6 TRANSPORT IN PERCOLATION AND RUNOFF WATERS

6.1 INTRODUCTION

(*a*) Soil Profile

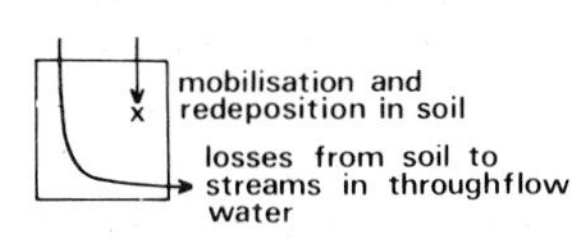

(*b*) Drainage Basins

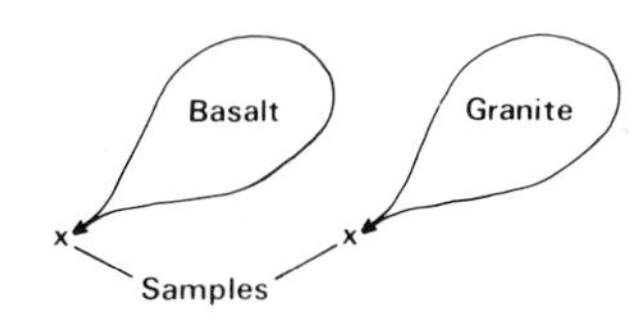

(*c*) Stream Transect

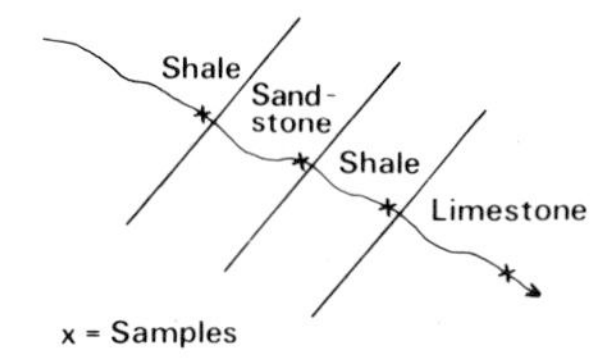

Figure 6.1 Solute investigation in geomorphological systems

This chapter considers the movement of solutes through soils and into streams and also the movement of sediment in streams. Transport of solutes through soils is a major factor influencing landform evolution, especially in humid temperate regions. It is also of interest in the applied context of the loss of plant nutrients, especially fertilisers. Once in streams the solutes derived from soils (sections 6.2–6.6) and the sediment derived from wash (5.3) and bank erosion (5.6–5.8) are transported in the body of water itself or along the bed of the stream. Transport is an important topic since it is a major limiting factor on the progression of erosion – frequently erosion can not proceed until weathered or previously eroded material is removed to reveal a fresh surface where new weathering and erosion can take place.

The principle points at which solutes can be investigated in geomorphological systems are shown in ***Figure 6.1***. Soil bodies can be investigated in the field by the study of solutes in vertically and horizontally moving water by the use of lysimeters (section 6.2 and throughflow troughs (6.3). Soil-water chemistry can be studied by the use of suction cups (6.4) and water flow in soils can be assessed by the use of soluble tablets (6.5). In the laboratory, leaching columns can provide much useful information about the soluble material in soil profiles (6.6). In streams (6.7) runoff from drainage basins of contrasting lithologies can be studied and the uptake of solutes as a stream crosses different lithologies may be monitored (***Figure 6.1***).

The particulate material moved in streams occupies a continuum of size ranges. The smaller particles (clay and fine silt sizes) are in ***suspension*** in the flowing water whilst larger particles (coarse silt and sand) are bounding or ***saltating*** along the river bed and the larger particles (cobbles and boulders) are dragged or rolled along the bed in floods as the ***traction load*** or ***bedload*** (sections 6.8–6.11).

6.2 LYSIMETERS

When infiltrating rain water moves down through the soil it may continue to percolate vertically down to the water table, especially in permeable soils and rocks. However, if permeability decreases downwards through the soil or bedrock then the water will tend to pond up and flow sideways downslope. Thus, water may move in a dominantly vertical or horizontal direction through the soil in accordance with the permeability of the soil, transporting soluble material towards ground water and to streams. The amounts of water and solutes moving in these directions can be studied by two principle methods. The downward moving percolation can be studied by the use of ***lysimeters,*** described below, and the laterally moving water, termed ***throughflow***, can be studied by the use of ***throughflow troughs*** (***see*** section 6.3).

6.2.1 Method

In principle, a lysimeter is a block of soil from which drainage waters can be collected. A simple device is illustrated in ***Figure 6.2***. A small lysimeter can be constructed from two tin cans, while larger devices may be made using plastic piping. Piping can be inserted

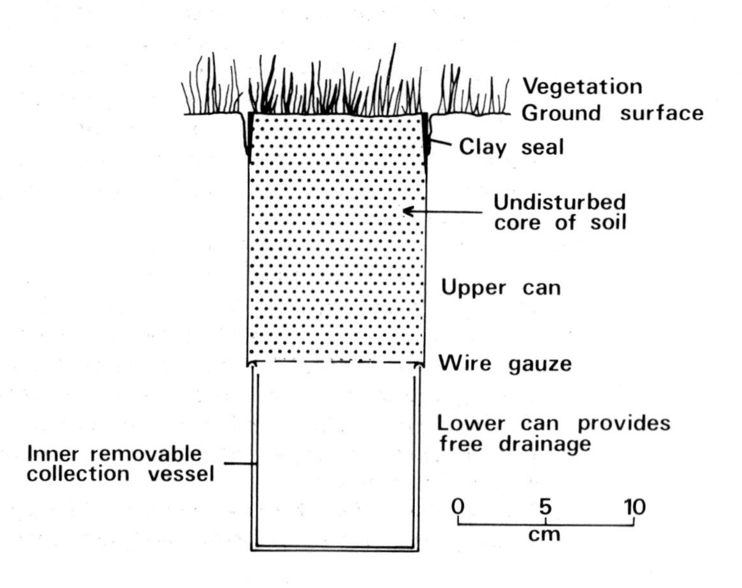

Figure 6.2 A simple lysimeter device made from two cans (After Curtis and Trudgill, 1975)

by simply hammering it into the ground, but this procedure is liable to markedly affect soil structure as both compression and shattering of aggregates may occur. Compression is undesirable since it will decrease flow rates and increase opportunities for chemical reaction above those encountered in an undisturbed soil. Rocking of the pipe and sideways movements increase cracking and this will lead to increased percolation rates and decreased chemical yields. It is thus important to preserve the soil structures as intact as possible and also to seal the sides of the pipe to prevent flow down the sides. As with an infiltrometer (section 5.2), it is better to have a large diameter pipe, as this will increase the area:perimeter ratio and decrease the edge-flow effects. The ***procedure*** for the insertion of a pipe or tube is as follows:

(1) Obtain a length of PVC tubing, of dimensions 15–30 cm diameter and saw into lengths appropriate to the soil depth. A 30 cm diameter, 30 cm length tube is appropriate, but a 15 or 20 cm diameter and 20 or 25 cm length tube is easier to handle. Wall thickness needs to be of the order of 1 mm or more. Bevel the lower edge to form a sharp edge. The length could be tailored to changes in the soil profile, for example if the soil 'A' horizon extends from 0 to 15 cm and the 'B' horizon from 15 to 40, two tubes could be cut, one 15 cm long and one 40 cm long, and inserted to assess the water and solute outputs from the 'A' horizon alone and the A + B horizon. Alternatively, several replicates of the 'A' horizon alone or the total profile depth could be made.

(2) Smear the inside of the tube with Vaseline, place the tube on the soil surface and lightly tap the top to mark out the position of the lysimeter. Inspect the circular mark and remove any large surface stones or twigs. Replace the tube and, holding the tube firmly in position, carefully excavate a trench round the ring, leaving a column of soil upstanding, of diameter slightly larger than the ring (***Figure 6.3***). During this process, remove any obstructions (roots may be sawn or cut off and stones removed or broken with a hammer and chisel). In a very stony soil, there will be considerable problems, but it is adequate either to push small stones slightly in or to replace large stones by smaller ones.

(3) Once a column of soil has been exposed, lightly tap the tube down the column, taking great care to keep it vertical and not to rock it from side to side.

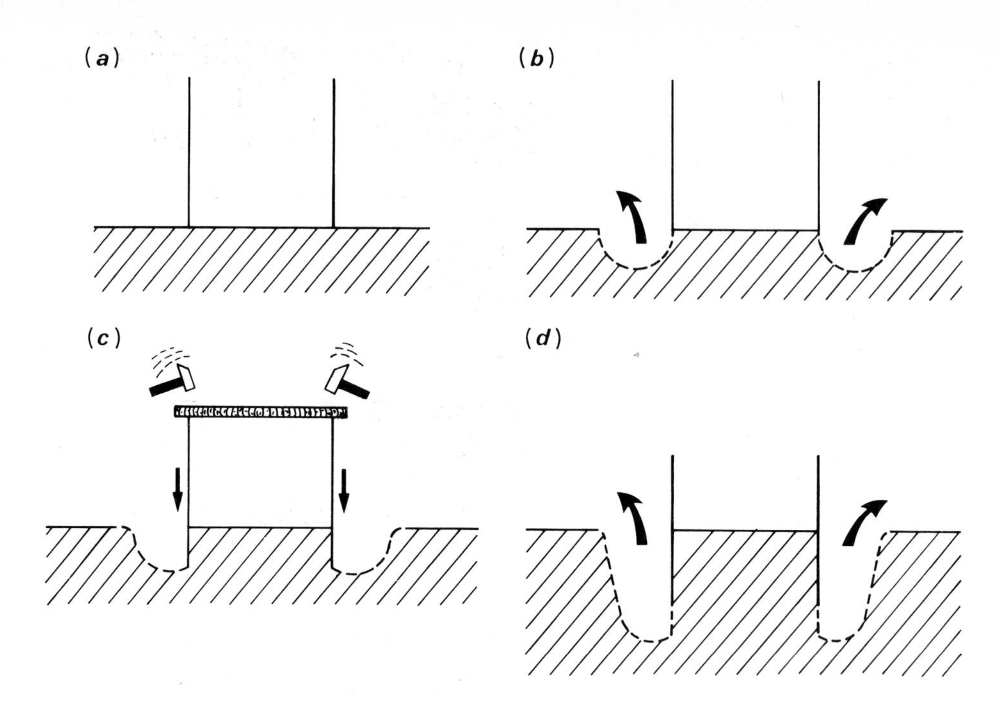

Figure 6.3 Insertion of lysimeter tube or leaching-column tube

(4) Repeat the excavation and lowering process until the tube is at the required depth and cut off the base of the column with a spade.

This procedure can be lengthy but is essential if compression and cracking are to be minimised. It is greatly preferable to hammering in which may cause the tubes to bounce on roots and lead to the break-up of soil structure. Stones cannot be removed during hammering and may be trapped under the leading edge of the tube. If this occurs and hammering is continued a substantial air space may form behind the stone as the tube is hammered in.

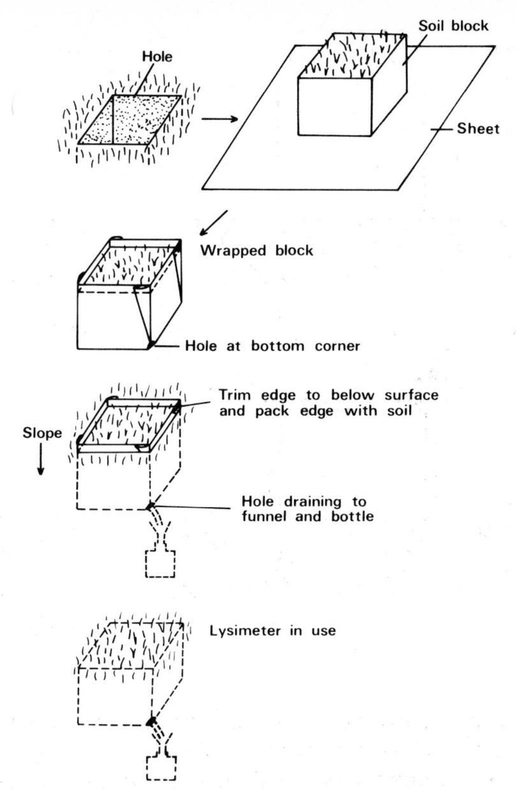

Figure 6.4 Installation of simple polythene-sheet lysimeter

Drainage is provided by excavating a pit below the site of emplacement and securing the column above the pit. The soil material can be prevented from falling out of the tube by the use of wire mesh with filter paper or muslin secured round the base. Drainage is collected using a funnel draining to a bottle below. The easiest way to achieve free drainage is to mount the soil-filled pipe on top of a second pipe, with gauze between. A funnel is fixed in the top of the lower pipe, with a cut-away section at the base of the pipe for access to the bottle.

The edge of the soil column should be sealed against the inner sides of the tube in order to minimise drainage down the sides. A locally available puddled clay will suffice, but kaolin clay can be purchased from chemists. Proprietary substances like Vaseline are also commonly used.

Alternatively, a soil column may be excavated and bounded by fibre glass and resin (as used for canoes). This minimises any edge flow, but is only successful on drier soils.

A considerably simpler, if crude, lysimeter can be made by digging out a spade-size block of soil intact and then carefully laying it back ***in situ*** on top of a polythene sheet folded to fit the hole. At one corner a second pit can be dug and a corner fold made into a tube flowing into a bottle (***Figure 6.4***). Most water will, however, flow down the sides rather than through the soil unless the sides are sealed in some way. Cutting the polythene below the surface and filling in with packed soil will minimise this problem.

6.2.2 Discussion

The lysimeter is left to drain naturally under rainfall, and note is made of the rainfall amount, the amount of drainage and the solute load – observations may be made on a per storm, day or week basis. The aim of the experiment might be to compare solute loads from contrasting soils under similar rainfall regimes, for example.

The study of Bourgeois and Lavkulich (1972) was designed to test the relationship between soil minerals and solute output from lysimeters and also the relationship between water flow and solute yields. They found that cation abundance in water withdrawn from lysimeters on forest soils was related to the abundance of the elements present in soil minerals, and the order of concentration in the water was Na $\geqslant$ Ca $>$ K $>$ Mg. They also recorded that flow and solute yields increased with increasing rainfall, indicating that the solute supply was high and not readily exhausted during the passage of percolation water.

During percolation from the base of a lysimeter, water saturation occurs before drainage can occur. This may lead to hydro-chemical processes not normally found at the base of the soil profile, especially in relation to the chemical reduction of ion species; thus, for example, nitrate may be reduced to nitrite if saturation is prolonged. Suction devices may be used to overcome this, drawing water out at water contents below saturation. The lysimeter is then termed a ***tension lysimeter***. This can be achieved by the use of a porous ceramic plate and evacuated chamber below the lysimeter or, more easily, using a suction cup (***see*** section 6.4) inserted into the side of the lysimeter. In this case the base of the lysimeter can be sealed with a plastic disc and epoxy resin and all water drawn out using the suction cup. In this way natural drainage is not replicated, but neither is it using a free drainage system. The choice of system thus depends on availability of equipment and it is important to interpret the data gained in terms of the limitations of the method used. If field study is not convenient, lysimeter cores can be returned to the laboratory for leaching studies (***see*** section 6.6).

6.3 THROUGHFLOW TROUGHS

The solute load of mobile soil waters can be studied by the use of throughflow troughs inserted in stream banks or in pits on hillslopes. In the former case, the amount of solutes derived from slopes and contributing to the stream solute load (section 6.7) can be assessed, and in the latter case the amounts of solutes derived from different slope positions may be assessed, provided that there is lateral throughflow occurring in the soil.

6.3.1 Method

A trough simply collects water draining from a section of soil. It can vary in sophistication from a biscuit tin sawn diagonally in half and rammed into a stream bank to a carefully excavated and installed set of plastic troughs with flow recorders. The former will probably collect water, but the greater the degree of sophistication, then the greater the degree of rigour with which the hydrological and chemical results can be interpreted. The trough should be 0.5–1.0 m long, shorter ones being easier to install (especially in stony soil) and longer ones collecting a more spatially representative sample of soil water. The lower portion of the trough cuts off a vertical section of soil; thus the trough collects all the water draining to the section along the trough. A simple design and installation is shown in ***Figure 6.5.*** Further methods are discussed by Atkinson (1978) and Knapp (1973). Sampling from the trough is normally via a drain hole in the corner of the trough, with a tube leading to a collection bottle. A simple overflow device can be

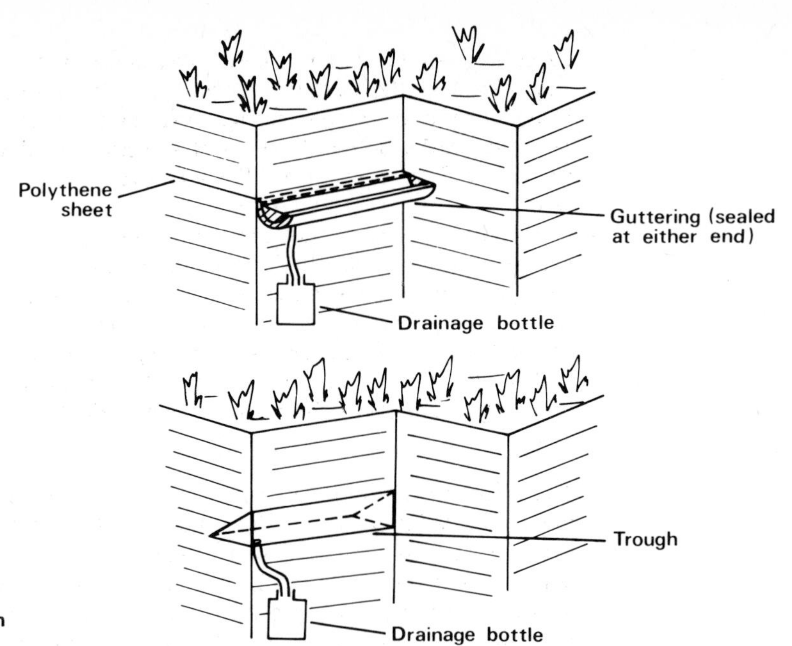

Figure 6.5 Throughflow trough installation

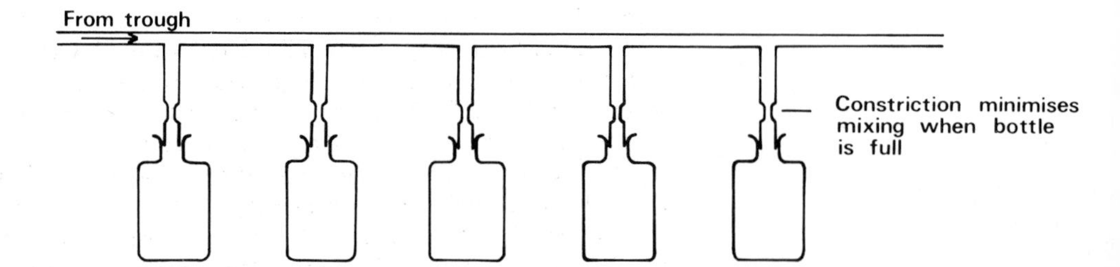

Figure 6.6 Bottle overflow collecting system

constructed so that earlier flows are collected separately from later flows (***Figure 6.6***) and, because of the narrow bottle necks, mixing from one bottle to the next is minimised. If more continuous records are required, the outlet can be connected to water samplers or recording rainfall gauges. In any of these cases, high flow may lead to the swamping of the system. Prior estimates of flow are therefore useful in indicating the most useful capacity of the collection system.

Probably the least disturbed site for a throughflow trough is a stream bank where a natural face already occurs. If, however, a trough is inserted in a slope site then the digging of a hole in which to insert the trough may cause the preferential flow of water to the trough. This problem can be partly overcome by placing the trough in the centre of a wider trench, or conversely backfilling the hole completely.

The trough itself can be backfilled with porous material to assist drainage, though, if this is done, care should be taken to use material that will not alter the chemistry of the flow.

6.3.2 Discussion

The experimental designs which can be involved are several and include (1) the monitoring of temporal variations in throughflow solute load over storms; (2) the relative losses of individual solutes at one site; (3) the variations in solute load and throughflow amount downslope or (4) in different topographic situations (e.g. hollows and spurs); (5) the contribution of hillslopes to streams; and (6) the comparison of troughs in similar topographic situations but (a) on different soils on similar bedrocks or (b) on different bedrocks.

Solute levels in waters draining from two contrasting soil types developed on Devonian Old Red Sandstone are shown in ***Table 6.1*** (after Waylen, 1979). The results suggest that brown earths produce a high solute yield, while podsolic soils, which are already leached, yield lower levels. From this study, Waylen deduced that, taking into account solutes and trough discharge over a one-year period, the podsolic area lost 3.1 tonnes of solutes km^{-2} a^{-1} while the brown-earth area lost 5.75 tonnes km^{-2} a^{-1}. Chemical evolution of drainage waters is shown in ***Table 6.2*** with data for rain, soil throughflow and runoff waters (after Smith and Dunne, 1977). Their work showed that most of the potassium uptake occurred in the soil, while much of the stream calcium and sodium was derived

TABLE 6.1 SOLUTE LEVELS IN THROUGHFLOW TROUGHS IN TWO SOILS ON OLD RED SANDSTONE (AFTER WAYLEN, 1979)

Soil horizon	pH	Mean chemical concentration of soil water (g ℓ⁻) Ca^{2+}	Mg^{2+}	Na^{+}	K^{+}	HCO_3^{-}	Cl^{-}	SO_4^{2-}	Si^{0}
					Brown earth				
A	4.80	2.37	2.30	3.37	12.26	2.29	9.55	15.08	2.34
B_1	5.06	3.29	2.47	6.92	12.70	3.42	16.45	25.97	4.57
B_2	4.97	2.40	3.18	8.51	4.63	2.49	17.36	15.39	3.73
B_3	5.63	3.16	2.90	5.21	8.55	2.12	17.58	10.25	1.81
C	4.30	4.30	4.75	8.17	10.78	1.45	24.03	18.53	4.12
					Peaty podsol				
A	3.75	2.24	1.22	3.42	2.46	0.00	6.23	11.78	1.12
B_{Fe}	4.02	1.51	1.35	4.11	1.14	0.73	5.73	9.70	1.52
B_2	4.09	1.77	1.00	3.43	0.94	0.47	5.54	9.32	1.23
B_3	4.28	1.34	1.47	5.46	0.47	0.15	7.26	10.68	1.83
C	3.90	2.19	1.20	4.86	0.65	0.14	7.46	10.38	1.46

TABLE 6.2 CHEMICAL EVOLUTION OF DRAINAGE WATERS (AFTER SMITH AND DUNNE, 1977), PODSOLIC SOILS WITH WEATHERABLE SILICATE MINERALS PRESENT

Source of solutes	Solute concentration (g ℓ^{-1}) SiO_2	Ca^{+}	K^{+}	Na^{+}
Rain water	0.0	0.8	0.5	0.3
Soil water	6.7	1.6	2.0	1.3
Stream water	9.2	1.6	0.9	1.3
Ground water	12.0	3.5	0.3	2.7

from the soil. Ground water contents of solutes were higher than soil or stream, especially for silica, because of the greater time for solute uptake, except for potassium which is strongly adsorbed onto soil clay surfaces.

Throughflow dynamics are illustrated in ***Figure 6.7*** which shows that while stream solute loads decreased during storm events, throughflow solute loads (conductance)

TABLE 6.3 SOLUTE LOSSES (Ca^{2+}) (mg ℓ^{-1}) FROM FIELD DRAINS (AFTER WILLIAMS, 1970)

(a) *Sandy soil*		Mean
1968	March	149
	April	167
	May	159
	June	129
	July	144
	August	151
	September	153
	October	163
	November	130
	December	144
1969	January	140
	February	147
	March	144
Mean 1968–69		148
(b) *Calcareous soil*		Mean
1968	March	222
	April	246
	May	245
	June	197
	July	246
	August	197
	September	152
	October	174
	November	165
	December	155
1969	January	133
	February	156
	March	210
Mean 1968–69		192

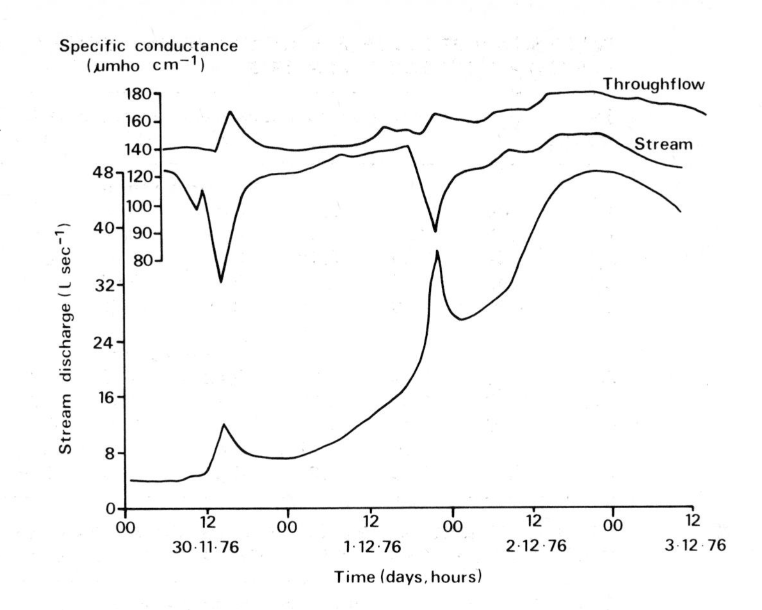

Figure 6.7 Throughflow dynamics on Old Red Sandstone (After Burt, 1979)

increased. This demonstrates that soil-water outflow of high concentration may exist during storm events, contributing to the solute load in runoff events. Soil water can thus be identified as a major source of solutes in a drainage basin (***see also*** section 6.6).

Tile drains can be used for sampling drainage waters from soil. These are often installed by farmers in wetter fields to facilitate drainage. Williams (1970) analysed the drainage from two soil types and the results are shown in ***Table 6.3.*** Seasonal effects are clearly visible as are the greater losses of Ca^{2+} from calcareous soils.

Soil-water solute outputs can thus be monitored in many ways and the data gained used to illustrate the processes of solutional denudation in soil-covered slopes.

6.4 SOIL-WATER SUCTION CUPS

Where drainage water is absent or in limited amount, throughflow troughs cannot be used effectively. Soil-water suction cups are then needed to gain a water sample.

6.4.1 Method

Porous ceramic suction cups (***see*** Appendix 1 for sources) are used to gain a sample of soil water by vacuum extraction. The method can be used to extract water held in the soil at tensions higher than field capacity (***see*** Briggs, 1977b, p. 66). The cup is inserted in an auger hole which is then backfilled with soil slurry. The cup is then evacuated by a hand pump and the sample is collected after 1–7 days (***Figure 6.8***). Chemical analysis of the sample can be used to demonstrate the solute concentration of soil waters, for example, down a slope profile.

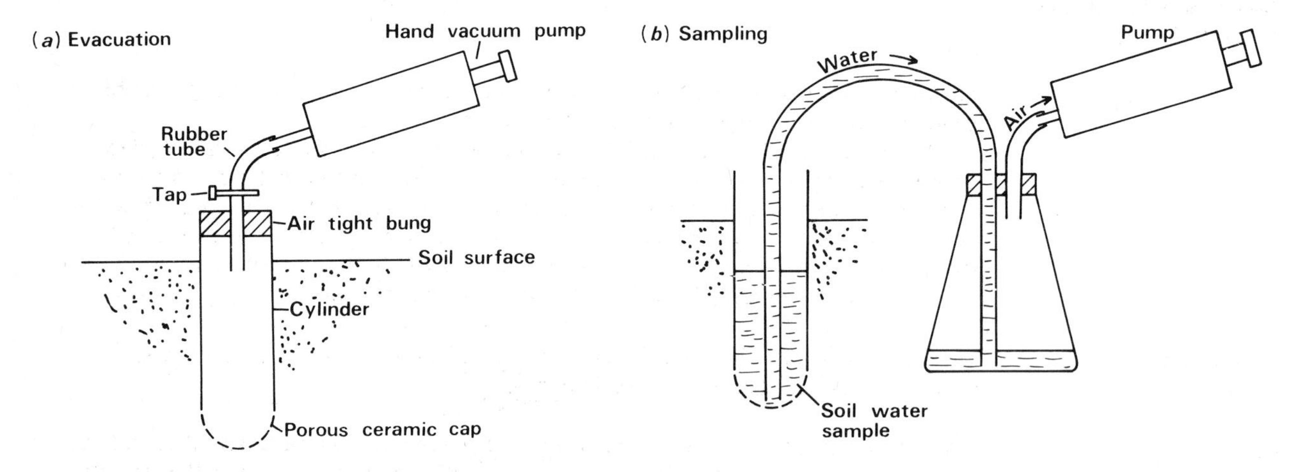

Figure 6.8 Operation of soil-water suction cups

6.4.2 Discussion

Figure 6.9 shows some sample values for downslope changes in soil water concentrations on a limestone hillslope, suggesting that solute concentrations increase downslope. The

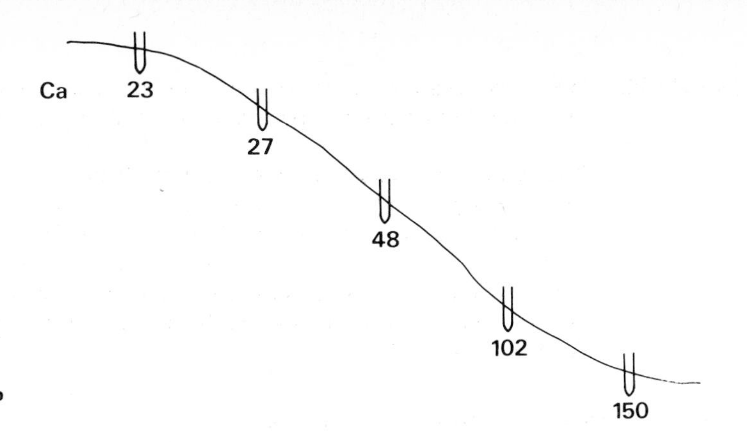

Figure 6.9 Sample values for suction-cup data on a limestone hillslope

suction-cup method is particularly useful on limestone hillslopes, and other permeable bedrocks, where the volume of throughflow waters is limited.

The presence of the cups may influence the values of soil-water solute levels gained because they draw water from differing areas of soil depending upon the soil moisture content. In addition, the porous ceramic may lead to some adsorption of ions. Thus, the detailed interpretation of the data may necessitate some qualifications, but provided standard procedures are adopted (especially method and depth of insertion and a standard vacuum) broadly interpretable results can be obtained.

6.5 Water flow in soils

The location of water flow in soils can be identified by the insertion and subsequent retrieval of soluble material in the soil profile.

6.5.1 Method

Gypsum, in plaster of paris form, dissolves in a manner that is independent of pH (at the ranges normally encountered in soils) but which is directly related to water flow rate. It follows that gypsum blocks can be inserted into a soil profile to identify any zone of rapid water flow in the soil. The blocks can be made from commercial plaster of paris and cast in a mould. As a sphere is a desirable shape (since it presents a uniform surface, without corners, to the environment) a mould for making ice-spheres can be used.

The spheres should be dried at 60°C for 24 h (not at a higher temperature since the gypsum will change to anhydrite), cooled in a desiccator and weighed to at least two decimal places of a gram. They can then be placed in the environment of interest and retrieved and reweighed after 1–6 months. Careful cleaning to remove dirt may be necessary.

6.5.2 Discussion

Figure 6.10 shows a soil profile on Carboniferous Limestone with the data for mass loss, demonstrating the location of water flow at the base of the soil profile. This is a clear

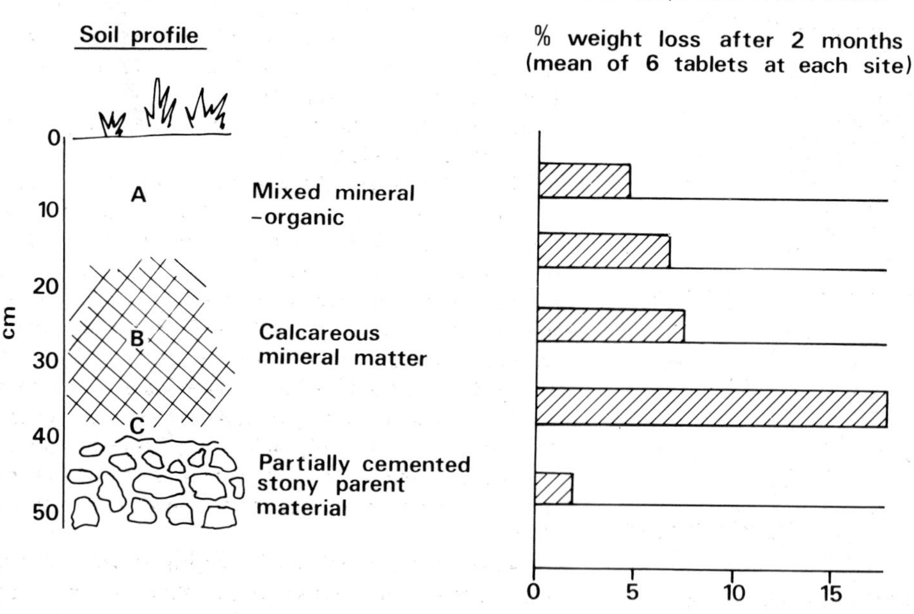

Figure 6.10 Water flow in a soil profile, using gyspum tablets

indication that in this case erosion by flowing water will be concentrated at 30–40 cm in the soil.

6.6 LEACHING COLUMNS

Water can be placed on top of a column of soil in the laboratory and the drainage or ***leachate*** examined to assess the amounts of readily soluble material which may be lost from the soil.

6.6.1 Method

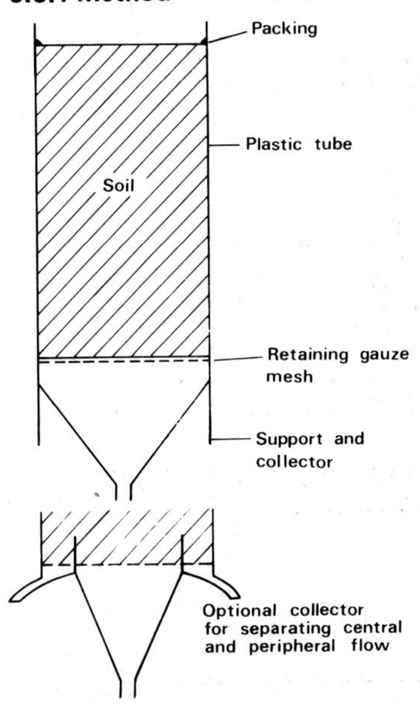

Figure 6.11 A leaching column

The principles of extraction of leaching columns from the field have been covered in section 6.2 (***Figure 6.3***) and should be adhered to in order to provide an undisturbed soil column for laboratory use. A soil column of a minimum length of 15–30 cm is probably the best size to aim for. Comparisons can also be made with columns from different levels (e.g. column 1 from the soil 'A' horizon alone, column 2 'A' + 'B', and column 3 to include weathering bedrock as well). In this way sources of solutes from different depths in the soil may be identified.

If edge flow down the sides of the soil column is a problem, packing at the top round the rim or a resin coat may help, as with lysimeters (section 6.2). Alternatively, it may be desirable to sample only the centre of the column and thus minimise the catch of any edge-flow water (***Figure 6.11***), though it should be emphasised that a total catch will not be made unless all the edge-flow water is also collected. The water may be funnelled into a bottle for analysis (section 3.4). This may be compared with the flow from another column. Alternatively the flow may be collected in a succession of several bottles changed over time if a temporal analysis of the leaching pattern is required.

Care should be taken over the quality of the water added at the top. It may be possible to collect rain water but, failing this, distilled water can be applied. This will usually be slightly acid (***see*** section 3.3) and will form an effective, if artificial, leaching agent. In addition, chelating compounds and organic acids may be used (***Figure 6.12***) (***see*** section 4.2).

As well as the problem of rapid leakage down the sides, one of the principle problems with leaching columns used in the laboratory is that of temperature control. Since the

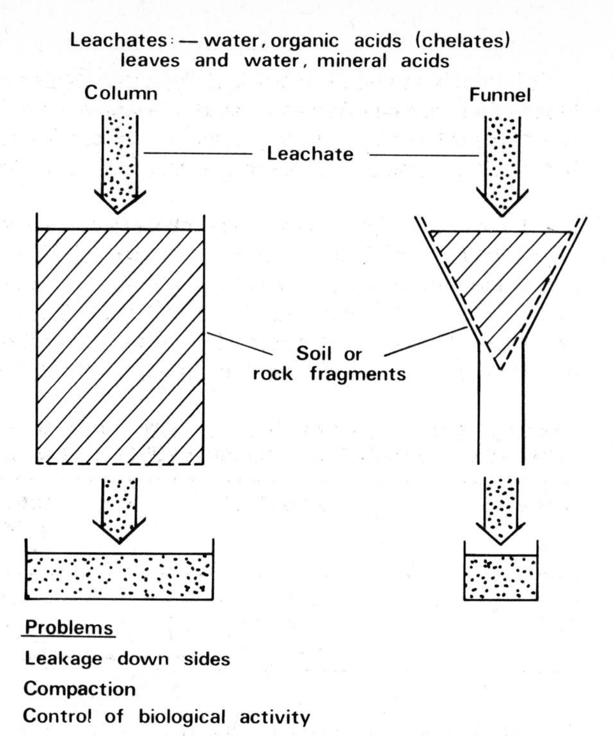

Figure 6.12 Leaching-column experiments

laboratory is usually warmer than the field situation it is likely that biological activity will be speeded up in the laboratory. This may affect the rates of biologically influenced chemical weathering (***see*** section 3.8). Thus the assessment of rates of operation of processes gained from such experiments may be difficult to transfer to the field situation. This will be especially true of nitrate as its mobilisation is very dependent upon biological activity. In this case, field study using lysimeters is preferable.

It is possible to build an artificial slope for leaching study in the laboratory; a large block of soil can be taken and placed on a board, bounded and sealed on the sides or it can be placed in a large tin tray. Water can be applied at the top and sampled at the bottom. The same problems apply to this method, however, as to the columns.

6.6.2 Discussion

Results for leaching columns using chelates have been obtained by Atkinson and Wright (1967) (*see* section 3.8), as shown in ***Figure 6.13***. These demonstrate the mobility of iron and aluminium with chelates. ***Table 6.4*** gives data for leaching of the A, B and C horizons of a calcareous soil, indicating that the C horizon's primary source of solutes is in the soil. Smith and Dunne (1977) passed 75 ℓ distilled water through a column over a nine-week period. The podsolic soil developed on a schist showed an initial flush of cations (Na^+, Ca^{2+}, K^+) which then maintained a low level in the leachate; SiO_2 levels were maintained, the silica being derived from the supply of weatherable silicate minerals, including muscovite, oligoclase, biotite and staurolite (*Figure 6.14*).

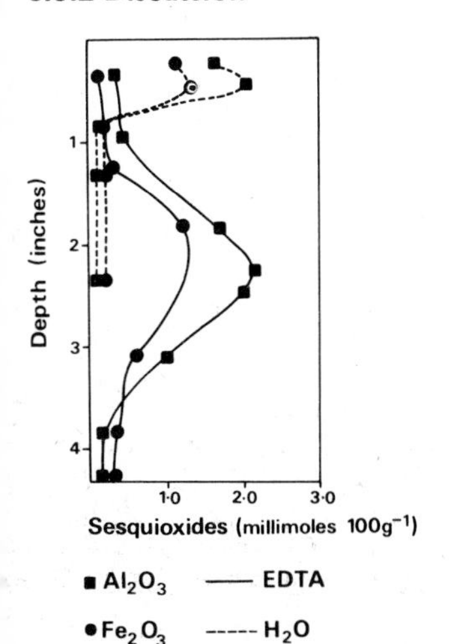

Figure 6.13 Leaching-column results using a chelate, EDTA (After Atkinson and Wright, 1967)

TABLE 6.4 SOURCE OF CALCIUM LEACHED FROM THREE LEACHING COLUMNS OF DIFFERENT HORIZONS IN A CALCAREOUS SOIL PROFILE

Horizon	% Solid $CaCO_3$	Calcium leached in distilled water (mg $ℓ^{-1}$)
A	0.2	2.3
B	3.5	32.3
C	17.2	102.1

6.7 SOLUTE LOAD IN STREAMS

The assessment of solute load in streams is a matter of taking a water sample from a stream and measuring the solute content (3.4) and working out the load in conjunction with discharge measurements.

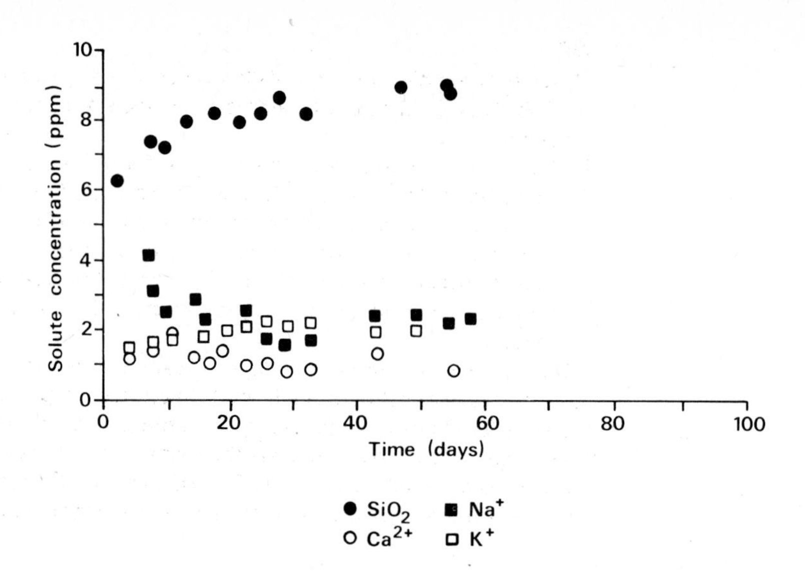

Figure 6.14 Leaching-column results, podsolic soil with silicate minerals (After Smith and Dunne, 1977)

6.7.1 Method

A suitable sampling bottle should be used; a screw-top polythene bottle of 25 ml size (which has been tested and found watertight) is normally ideal. In order to gain a representative sample, samples are best taken from a free-running turbulent section of the stream where the waters are well mixed. Backwaters, quiet eddies and sites immediately downstream of tributary streams or pipes should be avoided (unless these local variations are under study). Fill the bottle to the top, preferably by squeezing much of the air out and then holding the bottle under water, letting it fill out and the excess air escape.

The analysis can be for individual elements (section 3.4) or for total dissolved solids (TDS). The latter usually involves evaporating the water and weighing the residue or the measurement of conductivity. For the TDS by evaporation the apparatus required is evaporating basins, a balance accurate to three decimal places, and a drying oven capable of maintaining 105 °C.

TDS

(1) Clean the basin, wash in distilled water several times and dry in an oven for 2–3 h at 105 °C until all trace of water is gone; inspect for signs of dirt and, if present, clean again.

(2) Place the basin in a desiccator to cool (for at least 1 h) and weigh accurately.

(3) Place the water sample in the basin and replace in the oven until evaporated. A 250 ml sample in a wide, shallow 250 ml basin will take around 4–5 h to evaporate.

(4) Cool the basin and dried contents in a desiccator for at least 1 h and reweigh. The mass of basin + contents minus the mass of the basin will give a value of total dissolved solids in mg per sample volume; calculate the total dissolved solids (TDS) in mg ℓ^{-1}.

Experimental designs can involve the examination of downstream changes in solute load, the study of solute loads from different rock types and of temporal changes over flood events. Remember, however, if the sample appears at all cloudy either to let it settle or to filter it first to remove the suspended sediment. Thus this measurement and that in 6.5 are usually usefully performed together. Filtration may cause some small loss of solutes on the filter paper so settling and decanting may be preferable.

TABLE 6.5 CONVERSION FACTORS: mg ℓ^{-1} × F_1 = meq ℓ^{-1} (AFTER HEM, 1970)

Element and ion species	F_I
Hydrogen carbonate (HCO_3^-)	0.01639
Calcium (Ca^{2+})	0.04990
Carbonate (CO_3^{2-})	0.03333
Chloride (Cl^-)	0.02821
Magnesium (Mg^{2+})	0.08226
Nitrate (NO_3^-)	0.01613
Phosphate (PO_4^{3-})	0.03159
Potassium (K^+)	0.02557
Sodium (Na^+)	0.04350
Sulphate (SO_4^{2-})	0.02082

The chemical analyses used will depend upon the availability of facilities and also the solutes likely to be present in the stream in question. Conductivity can be used as a useful general indication of solute concentrations. However, while individual analyses – for example of calcium in limestone streams – are useful, calculation of total ionic balance can be a valuable check on the methods and also indicate the overall composition of the water. Here the results in mg ℓ^{-1} are multiplied by the factors given in ***Table 6.5*** to convert to milli-equivalents per litre (meq ℓ^{-1}). When the charges on the cations and anions are added together they should balance. If they do not there is either some error in analysis or a substantial amount present of some ion not analysed.

6.7.2 Discussion

Load can be calculated by multiplication of discharge by solute load. Discharge measurement is outside the scope of this book but reference can be made to Hanwell and Newson (1973), Smith and Stopp (1978) and to Goudie (1981) for further techniques. Load calculations are important since, while concentrations may fall at high water flow, there may be so much water present that the actual amount removed (the load) may increase. Thus, for a low flow, a river might have a solute concentration of 10 mg ℓ^{-1} and a discharge of 5 ℓ s^{-1}, giving a load of 10 X 5 = 50 mg s^{-1}; in flood, the concentration may only be 2 mg ℓ^{-1}, but the flow may be as high as 100 ℓ s^{-1}, giving a much higher load of 2 X 100 = 200 mg s^{-1}.

Solute budgets for some Norfolk, UK, river catchments have been studied by Edwards (1973), who derived values for annual losses of solutes on the basis of tonnes per square kilometre. Losses of 20–30 tonnes km^{-2} a^{-1} for calcium were found in an area of calcareous boulder clay; by comparison, values for other cations ranged from 0.9 to 7.3 tonnes km^{-2} a^{-1}. Walling and Webb (1978) have related gross annual solute load in a drainage basin (The River Exe, UK), ***Figure 6.15a***, to underlying geology (***Figure 6.15b***), high solute loads draining from calcareous bedrocks.

Downstream changes in water chemistry are shown in ***Figure 6.16*** for calcium in a limestone stream, illustrating the uptake of calcium from the stream bed and soils of the stream bank; the rate varies but has a mean of 0.25 mg ℓ^{-1} Ca^{2+} m^{-1}; pH also increases in parallel to Ca^{2+} as H^{+} ions are used during hydrolysis (***see*** section 3.3).

Data presented by Hem (1970) are shown in ***Figure 6.17*** for solute concentrations in

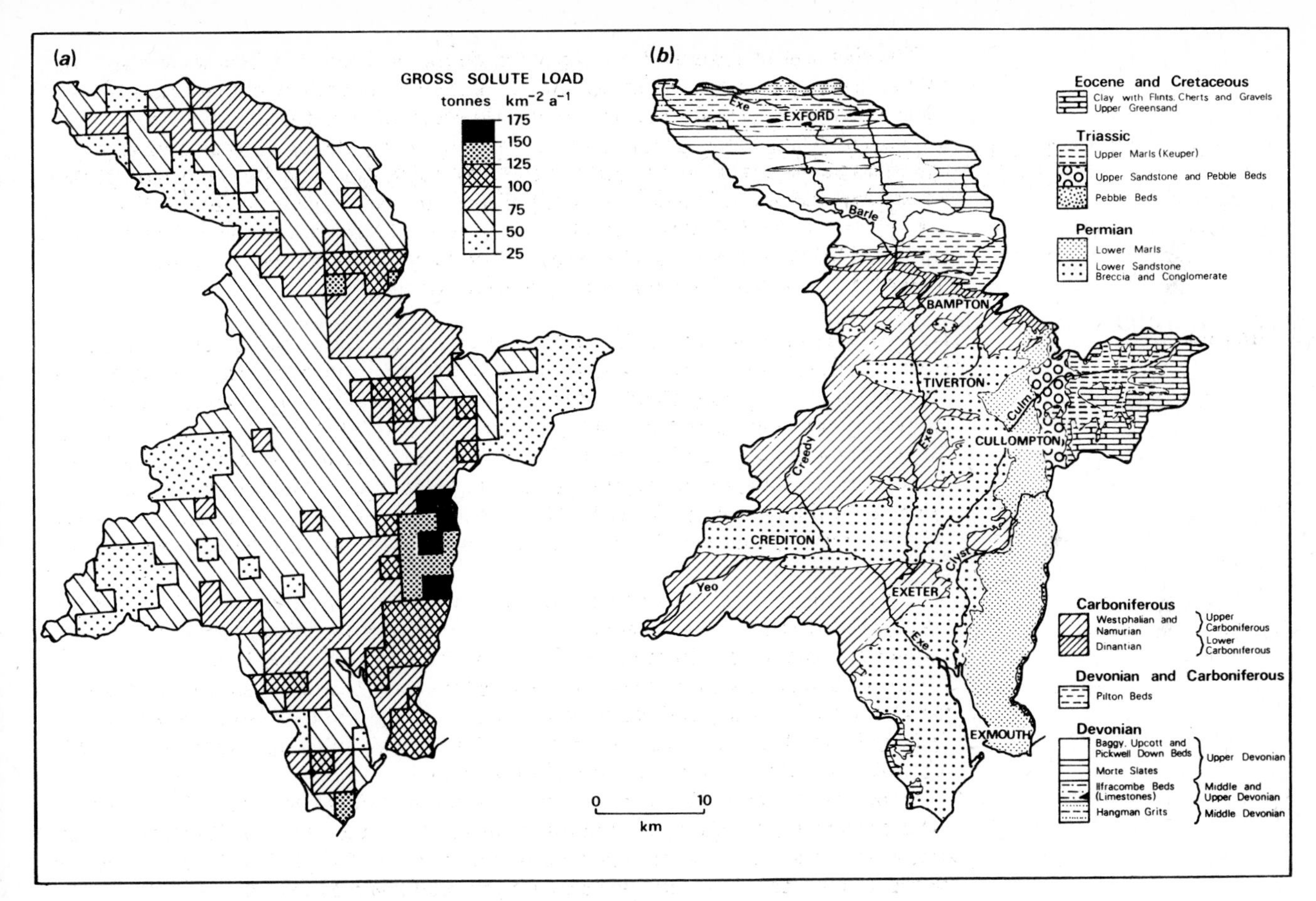

Figure 6.15(*a*) Gross solute load, Exe basin, and (*b*) geology, Exe basin

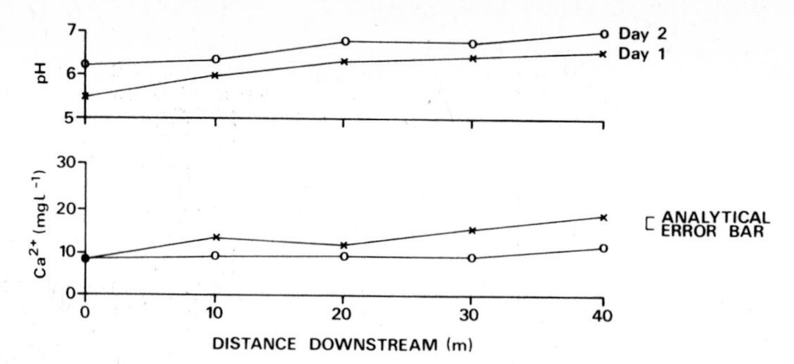

Figure 6.16 Downstream changes in Ca^{2+} concentration in a limestone stream (Rough Sike, Moor House, N. England) (After Trudgill, 1979)

TABLE 6.6 SOLUTIONAL DENUDATION RATES COMPUTED FROM LONG-TERM RECORDS (AFTER YOUNG, 1974)

Method	Climate	Rock	Ground lowering Limestone B*	Other B*	Location
Carbonate in rivers, world summary	Various	Limestones	25–100	–	World
Silica in rivers, world summary	Various	Various	–	1–6	World
Dissolved river load, USA summary	At 1000 mm a^{-1} rainfall	Various	–	30	USA
	At 500 mm a^{-1} rainfall	Various	–	2–15	USA
Dissolved river load	Polar maritime	Metamorphic	–	5–10	Scandinavia
Dissolved river load	Temperate continental	Sandstones	–	26	Polish Carpathians
Dissolved river load	Temperate continental	Till	–	30	New Hampshire, USA
Dissolved river load	Temperate continental	Limestone; various	32	18	Poland
Dissolved river load	Temperate continental	Sedimentaries	–	2	Maryland, USA
Dissolved river load	Mediterranean	Dolomite; granite	17–21	1–2	California, USA
Dissolved matter in slope wash	Temperate montane	Limestone	28	–	Austrian Alps
Dissolved matter in soil solution	Temperate maritime	Chalk	17	–	Southern England
Theoretical calculation	At 10° C, 1000 mm a^{-1} rainfall	Limestone	50	5	

*See Table 5.2, pp. 125–126

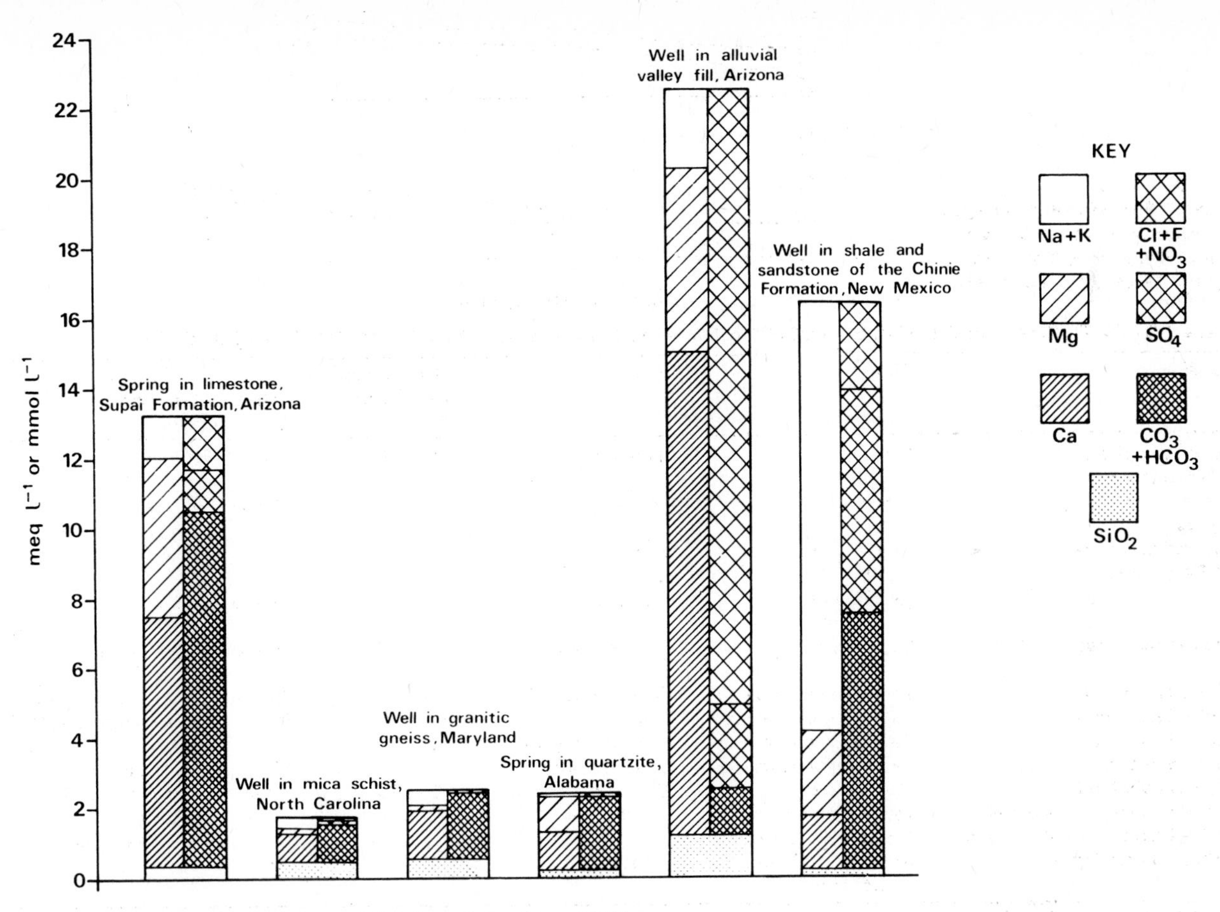

Figure 6.17 Solute concentration and lithology (After Hem, 1970)

relation to lithology, demonstrating the range of silica and cation values in relation to sedimentary and non-sedimentary rock types. ***Table 6.6*** shows the compilation by Young (1974) of solutional denudation computed from long-term studies of solute loads.

6.8 SUSPENDED SEDIMENT IN STREAMS

Measurement of sediment suspended in stream water will indicate how much a river is carrying. The importance of mechanical erosion can also be assessed.

6.8.1 Method

Sampling can be a major problem since sampling with a hand-held bottle is inadequate as this will disturb the flow of sediment. A simple sediment-sampling device which causes the minimum of disturbance is shown in ***Figure 6.18***. When collected, the analysis is simply a matter of weighing a dry filter paper, filtering the sample, drying the paper at around 50–60 °C for 1–2 h (i.e. so that it is dry but not burnt), cooling it in a desiccator for 1 h and then reweighing in order to express the mass of sediment as mg ℓ^{-1} suspended sediment. In addition, the sediment can be examined to see if it is mineral or organic in

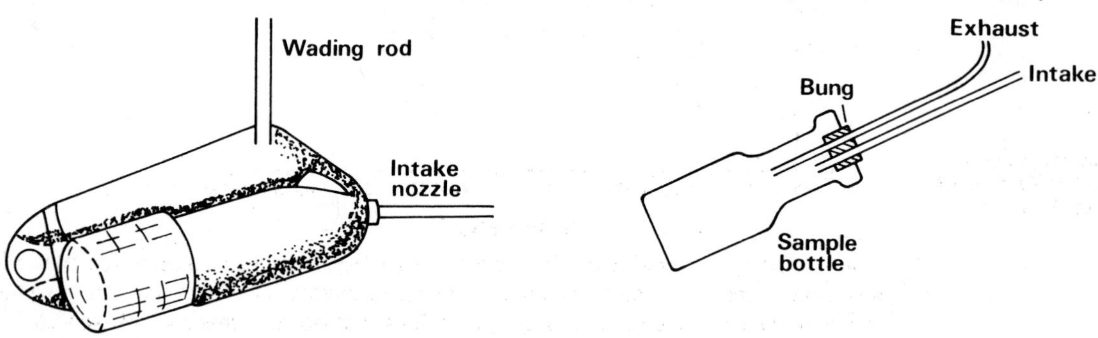

Figure 6.18 Suspended-sediment sampling (From Smith and Stopp, 1978)

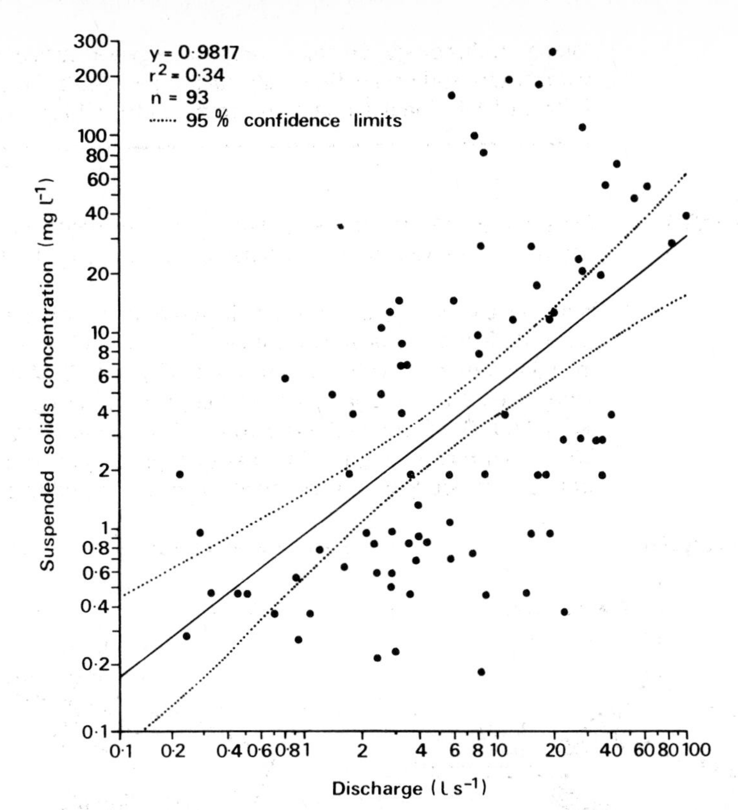

Figure 6.19 Suspended sediment–discharge relation for East Twin Brook, Mendip Hills, Somerset (From Finlayson, 1978)

nature (hydrogen peroxide can be used in order to oxidise organic matter and thus mass loss can be used to measure the amount of organic matter present). Again, measurements can be designed to show, for example, variations over storm events and differences between different catchments and the load can be calculated as in section 6.7. An example of a sediment–discharge relation is given in ***Figure 6.19***. The scatter of the data

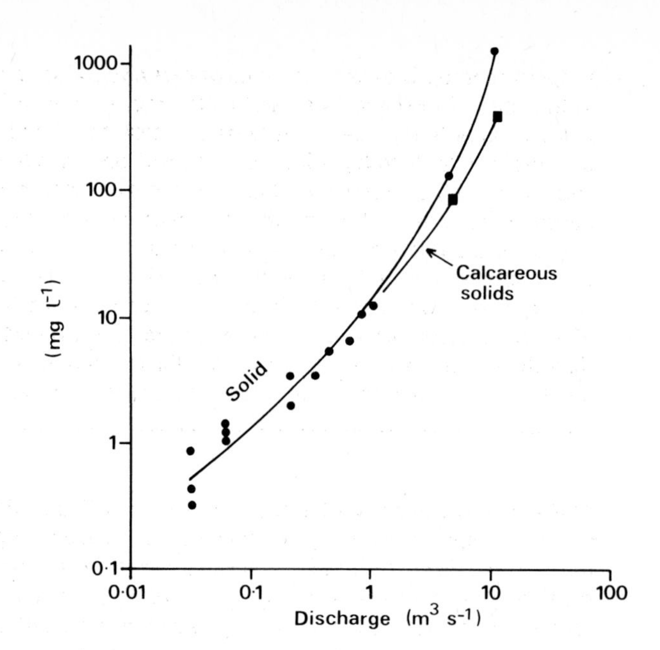

Figure 6.20 Suspended-sediment transport and discharge (From Newson, 1971)

is high since falling discharge and rising discharge frequently have contrasting sediment concentrations, much of the sediment being initially derived from the start of the storm; alternatively bank collapse at high flow may lead to increased sediment pulses. However, sampling over a range of flows will usually show that most of the sediment transport is achieved by high flows as, for example, shown by Newson (1971) (*Figure 6.20*).

6.9 BEDLOAD COMPETENCE

Movement of material on the stream bed depends upon the ability of the stream to carry particles of different sizes. This can be simply illustrated by taking a range of particles on a standard surface and seeing which sizes can be swept off the surface by the stream. The simplest method is to select a range of particles and, starting with the smallest, place each in the palm of the hand and insert it flat into the stream flow. Do this for each particle in turn and note the size of the largest particle that is moved. The method can be made to be more standardised by using a board with coarse sandpaper mounted on the upper side. Use material of the same rock type and similar shapes and place the board at a standard depth in the stream. Hold the material in place with a finger until the board is in position. Withdraw the finger and observe whether the particle is moved. It is of course possible to vary the rock type while holding size and shape constant and to vary shape while holding rock type and size constant. Thus, it can be ascertained if there is any variation in the effects of the combinations of these factors. It is difficult to translate these results directly into the natural situation since the method in no way replicates natural conditions; its use is in illustrating the principles involved.

6.10 BEDLOAD TRAPS

The movement of bedload is one of the most difficult things to measure in the study of erosion, simply because any device which is introduced into the system to measure the movement is liable to influence the operation of the process. This is, of course, true for many other measurements but it is particularly so in this case as most of the methods tend to involve the use of a trap into which material falls and it is difficult to assess how far the presence of the trap – usually some form of hole in the bed of the river – will encourage the movement of pebbles when they would not have otherwise moved. In the natural state bedload particles are held against each other by friction and these friction forces are removed at the lip of the bedload trap. Thus, during medium or high flows bedload may move into the trap when the bedload itself is not otherwise moving.

6.10.1 Method

A standard type of bedload trap which suffers from the kind of problems outlined above is illustrated in ***Figure 6.21a.*** Ignoring, for a moment, the problems outlined above, it is evident that the amount of bedload can be weighed and the total mass can be compared with the discharge of the river; it may thus be possible to derive a graph for bedload–discharge relations. It is also possible to sieve the sediment (Briggs, 1977a, pp. 64–68)

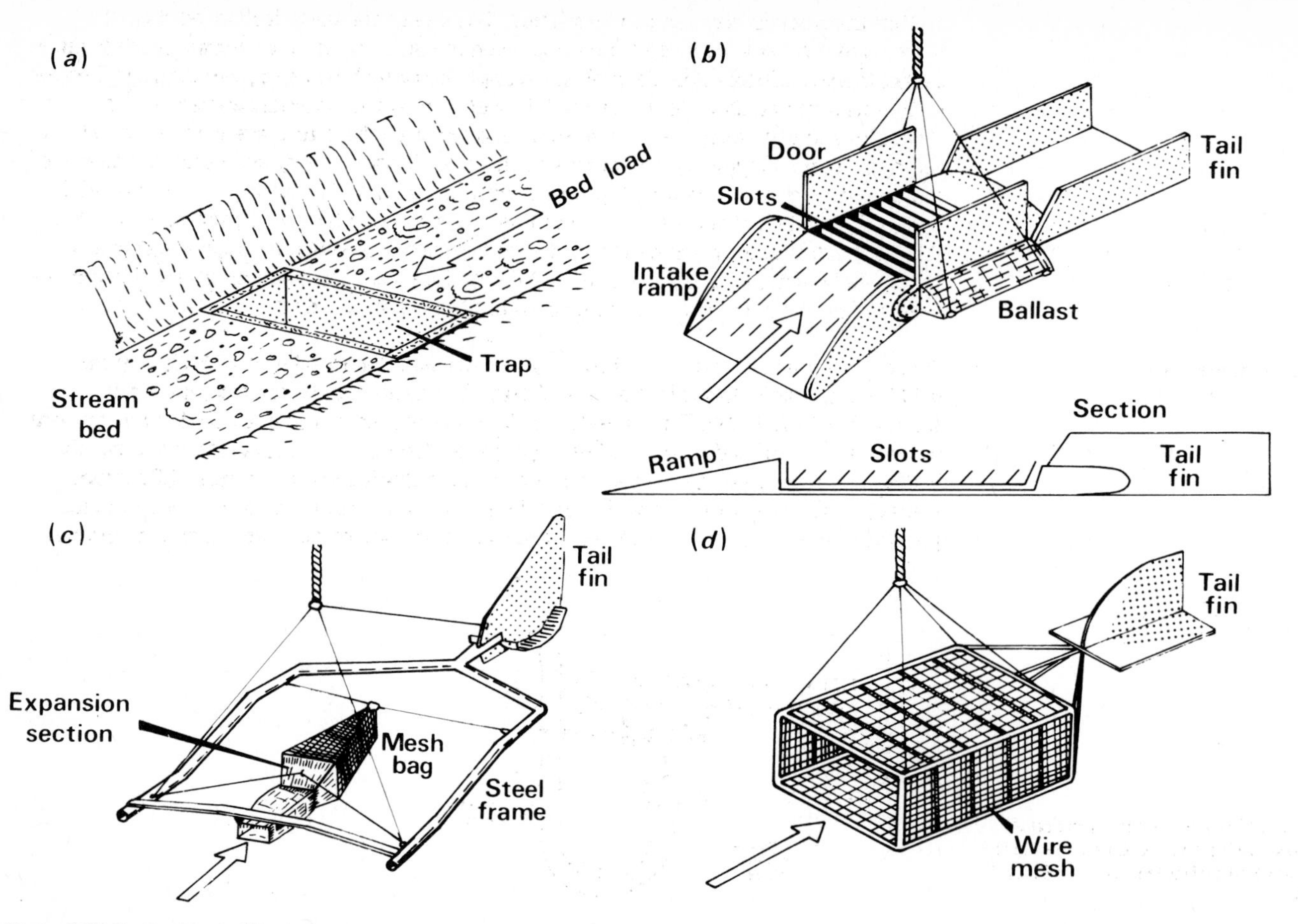

Figure 6.21 Bedload traps (From Gregory and Walling, 1973)

so that the size distribution can be studied. The size of the particles moved should increase as discharge increases; the force being required to move, or ***entrain***, each particle size is shown in Briggs (1977a, p. 95). An apparatus which helps to overcome some of the criticisms outlined above is a wire-mesh basket into which material can pass at the upstream end but which has a mesh small enough at the base to prevent the material falling through. Obviously, the mesh size chosen will have to vary according to the size of material it is wished to study. A problem with this method is that very coarse material may lodge at the entrance and that the finest material will fall through the bottom. If a solid bottom is used it will be found that material may be swept through the device as there will be little friction. Experimentation on individual streams will reveal which is the most efficient device for each situation of prevalent bedload sizes.

6.10.2 Discussion

A variety of bedload traps are shown by Gregory and Walling (1973, p. 160) (***Figure 6.21***) and a simple device is shown by Welton and Ladle (1979) for trapping small sediment (***Figure 6.22***). This is useful for low flows as most studies of bedload movement show that the bedload is immobile except under flood conditions. A relation between bedload size and velocity needed to move the particles is shown in ***Figure 6.23*** (from Helley, 1969); the slope of the line and data points vary from site to site according to particle shape and density but the principles of entrainment discussed above remain the

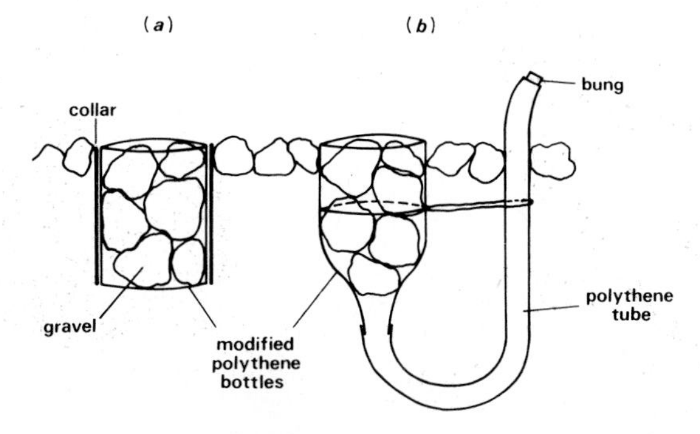

Figure 6.22 Device for trapping small particles moving across larger pebbles in a stream bed (After Welton and Ladle, 1979)

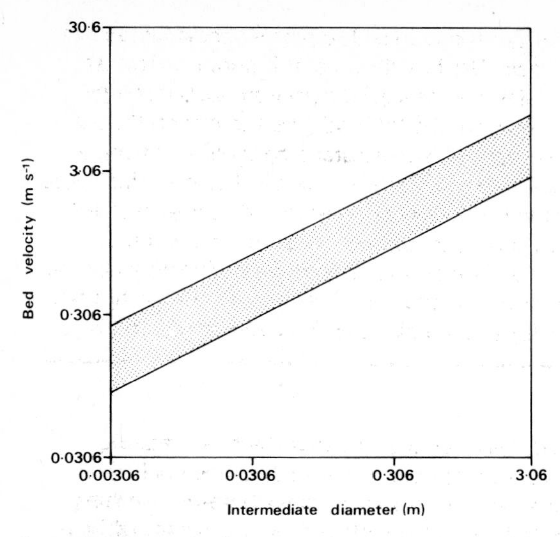

Figure 6.23 Bed velocity and size of particle moved (After Helley, 1969)

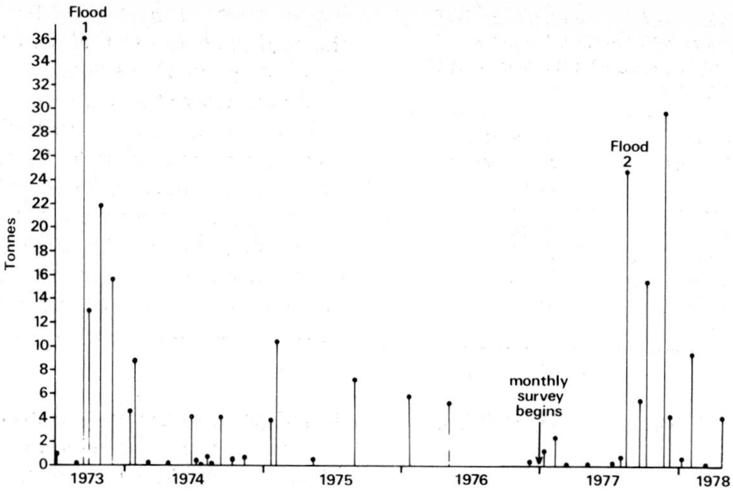

Figure 6.24 Bedload-trap yields, mid-Wales (After Newson, 1980)

same for all streams. Newson (1980) relates bedload movement to flood events in mid-Wales (***Figure 6.24***) with over 20 tonnes being removed from one bedload trap after a flood in August, 1977. This demonstrates the importance of infrequent, major events in moving large amounts of sediment.

6.11 BEDLOAD MARKING

A simple method, which can often be usefully used in conjunction with section 6.10, is that of painting pebbles and following their movement downstream. The choice of paint is crucial and a hard-wearing yacht-hull paint is usually satisfactory and often survives over one season. In making a choice of colours it is necessary to bear in mind the natural

TABLE 6.7 MOVEMENT OF BED-LOAD AFTER ONE YEAR'S EMPLACEMENT IN AN UPLAND STREAM

Size (mm) ('b' axis)	Movement downstream (m)
1–2	2–50
2–5	2–50
5–10	2–10
10–50	0–3
50–100	0–1

inquisitiveness of people and the popularity of the site and to balance these factors with the need to be able to find the pebbles again. Bright yellow paint is poor on the first count but good on the latter – a green paint may be a good compromise, but several trials exist where the operators have returned to find their pebbles laid out on the bank by someone curious as to their distinctive nature! Notwithstanding these problems, a useful approach is to grade the pebbles as to size or shape or both and to mark individual classes with different colours. An hypothesis can then be generated that smaller sizes should travel furthest and that the largest stones should only move during extreme events; similarly, rounded stones may travel further than angular ones. The effect of the paint is perhaps to make the pebbles smoother so that they travel a little faster, but the effect is liable to be small. Some results for a mountain stream are shown in ***Table 6.7***.

6.12 GENERAL DISCUSSION

A final point that should be made is that it is often the case that the most valuable projects are those which use two or more techniques together in the same field area. Thus, for example, solute movement may be first predicted using laboratory leaching columns and then theories tested in the field by the use of lysimeters and throughflow troughs together with the study of stream solute loads. In this way an integrated pattern of overall processes may be established.

APPENDIX 1 Sources (mostly UK)

CARBON DIOXIDE	Portable CO_2 measurement probe. (Gastec) D.A. Pitman Ltd, Jessamy Road, Weybridge, Surrey KT13 8LE.
CHEMICAL SUPPLIES	Including test kits and colour comparators. BDH Ltd, Poole, Dorset BH12 4NN.
CONDUCTIVITY & pH METERS FOR FIELD USE	WPA Ltd, Shire Hill, Safrron Walden, Essex CB11 3BD.
COURSES: FIELD STUDIES COUNCIL	Course on methods and syllabus material for sixth formers: Education Officer, Field Studies Council, Preston Montford, Montford Bridge, Shrewsbury, Shropshire S74 1HW (0743 850674).
EQUIPMENT	Gallenkamp, P.O. Box 290, Technico House, Christopher Street, London EC2P 2ER.
SCHMIDT HAMMER	Engineering Laboratory Equipment, Frogmore Road, Hemel Hempstead, Herts. HP3 9RL.
SUCTION CUPS	Soil moisture equipment corp., Santa Barbara, P.O. Box 30025, California CA 93105, USA.
TECHNICAL INFORMATION	***British Geomorphological Research Group, Technical Bulletins,*** published by Geo Abstracts, University of East Anglia, Norwich NR4 7TJ.
WATER ANALYSIS	BDH water hardness reagents from British Drug Houses Ltd., Laboratory Chemicals Division, Poole, Dorset, BH12 4NN.
	HACH Test kits for water analysis from Camlab Ltd, Nuffield Road, Cambridge CB4 1TH.

APPENDIX 2 Safety

(1) Never use any chemicals before checking bottle labels for safety hazards. Use protective clothing and fume cupboards when necessary. Check and abide by laboratory rules in operation. ***If in doubt, do not use chemicals.*** Dispose of all chemicals safely.

(2) Always work under supervision in a laboratory: do not work alone.

(3) Do not work alone in the field unless forced to: if you do, leave a note of route and return time with someone who will give an alert if you fail to return. Cancel the note on your return. Wear warm and waterproof clothing and strong footwear.

(4) Do not work under unstable vertical surfaces such as loose cliffs or in trenches. Always wear a helmet when under rock faces, quarry faces or cliffs of any kind.

(5) Wear goggles when using a geological hammer; do not hit one geological hammer with another or a chisel as the hammer may splinter (use a lump hammer with a chisel).

(6) Seek permission for ***all*** field visits, especially quarries; active quarries are extremely dangerous and should not be visited unless being guided by a representative of the quarry owner.

(7) Beware traffic on road sections; avoid where possible.

(8) Beware of danger of drowning when water sampling; avoid sampling from rivers in flood unless using a safe position, roped if necessary.

GLOSSARY

ABNEY LEVEL	A device for measuring slope angle.
ABSORPTION	Sorption within pores of a medium.
ACID	An acid dissociates to yield H^+ ions.
ADSORPTION	Sorption on to a surface, often of a cation (positive ion) to a negatively charged surface.
AGGRESSIVENESS	Ability of water to dissolve limestone.
ALIQUOT	A subsample of identical properties to another.
BEDLOAD	That part of the river load of particles moving at or on the river bed.
CHELATION	The incorporation of a metal, such as calcium, into the chemical structure of an organic compound such as an organic acid.
CONDUCTANCE	The ability of water to conduct an electrical current, measured in mhos and dependent on total ionised solute content.
DESORPTION	Detachment of a charged ion from an adsorption site.
Eh	Oxidation—reduction potential; positive values for oxidation, negative for reduction, ***see Figure 3.7***.
ENTRAINMENT	Dislodgement from the rest position and transportation in moving water, air or ice.
FLOCCULATION	Aggregation of clay particles, usually by chemical means.
$[H^+]$	Square brackets indicate concentration.
HYDRATION	Uptake of water by a mineral.

HYDROLYSIS	Water may dissociate to yield H^+ and OH^-; the reaction of a mineral with these constituents of water is termed hydrolysis.
IGNEOUS ROCKS	Those laid down by primary volcanic action.
KARST WATER	Water flowing through a karst (eroded limestone) system.
LITHOLOGY	Relating to rock type.
LYSIMETER	Field soil column for soil moisture and drainage studies.
METAMORPHIC ROCKS	Those changed by heat or pressure from their original state.
OXIDATION	Loss of an electron; iron(**II**) is oxidised to iron(**III**).
pH	Acidity or alkalinity; $pH = -\log_{10} [H^+]$; low pH = high acidity.
POROSITY	The total pore space in a soil or rock.
REDUCTION	Gain of an electron; iron(**III**) is reduced to iron(**II**).
SALTATION	A jumping action exhibited by smaller particles too large to suspend but smaller than bedload.
SEDIMENTARY ROCKS	Those laid down after erosion and transportation or biogenically.
SOIL HORIZON	An horizontal layer in a soil profile, differentiated from lower and upper layers by colour, texture, stoniness or other characteristic.
SOLUBILITY	Solids can dissolve in aqueous media. Their concentration at equilibrium defines their solubility.
SWALLET	Point of engulfment of stream in a limestone area.

REFERENCES

Adams, W.A., Evans, L.J. and Abdulla, H.H. (1971) 'Quantitative pedological studies on soils derived from Silurian mudstones. III. Laboratory and ***in situ*** weathering of chlorite. ***Journal of Soil Science*** **22**, 158–165

Allen, S.A. (Ed.) (1974) ***Chemical analysis of ecological material.*** London, Blackwell

Atkinson, T.C. (1978) 'Techniques of measuring subsurface flow on hill slopes.' In Kirkby, M.J. (Ed.) ***Hillslope hydrology***. Chichester, Wiley

Atkinson, H.J. and Wright, J.R. (1967) 'Chelation and vertical movement of soil constituents.' In Drew, J.U. (Ed.) ***Selected papers in soil formation and classification***. Soil Science Society of America, Special publication, **1**, 326–335

BDH (1973) ***Chemical methods of water testing.*** BDH chemicals, Poole, Dorset, UK

Birkeland, P.W. (1974) ***Pedology, weathering and geomorphological research***. Oxford, Oxford University Press

Blackwelder, E. (1933) The insolation hypothesis of rock weathering. ***American Journal of Science*** **26**, 97–113

Bloomfield, C. (1953a) 'A study of podsolisation. I. The mobilisation of iron and aluminium by Scots Pine needles'. ***Journal of Soil Science*** **4**, 5–16

Bloomfield, C. (1953b) 'A study of podsolisation. II. The mobilisation of iron and aluminium by the leaves and bark of ***Agathis australis*** (Kauri).' ***Journal of Soil Science*** **4**, 17–23

Bloomfield, C. (1954a) 'A study of podsolisation. III. The mobilisation of iron and aluminium by Rimu (***Dacrydium cupressinum***).' ***Journal of Soil Science*** **5**, 39–45

Bloomfield, C. (1954b) 'A study of podsolisation. IV. The mobilisation of iron and aluminium by picked and fallen larch needles.' ***Journal of Soil Science*** **5**, 46–49

Bloomfield, C. (1954c) 'A study of podsolisation. V. The mobilisation of iron and aluminium by Aspen and Ash leaves.' ***Journal of Soil Science*** **5**, 59

Bollinne, M. (1976) 'The experimental station at Sauveniene, Gembloux, Belgium.' In De Ploey, J. (Ed.) ***First Benelux Colloquium on Geomorphological Processes, Louvain, 1976***. (Abstracts of papers)

Bourgeois, W.W. and Lavkulich, L.H. (1972) 'A study of forest soils and leachates on sloping topography using a tension lysimeter.' ***Canadian Journal of Soil Science*** **52**, 375–391

Briggs, D.J. (1977a) ***Sediments***. Sources and Methods in Geography, London, Butterworths

Briggs, D.J. (1977b) ***Soils***. Sources and Methods in Geography, London, Butterworths

Buchan, G. and Ritchie, W (1979) 'Aberdeen Beach and Donmouth Spit: an example of short-term coastal dynamics.' ***Scottish Geographical Magazine*** **95**, 27–43

Burt, T.P. (1978) ***Three simple and low-cost instruments for the measurement of soil moisture properties***. Huddersfield Polytechnic, Department of Geography and Geology, Occasional Paper No. 6, 30 pp.

Burt, T.P. (1979) 'The relationship between throughflow generation and the solute concentration of soil and stream water.' ***Earth Surface Processes*** **4**, 257–266

Carrara, P.E. and Carroll, T.R. (1979) The determination of erosion rates from exposed tree roots in the Piceance Basin, Colorado. ***Earth Surface Processes*** **4**, 307–317

Cooke, R.U. (1979) Laboratory simulation of salt weathering processes in arid environments. ***Earth Surface Processes*** **4**, 347–359

Curtis, C.D. (1976) Stability of minerals in surface weathering reactions: a general thermochemical approach. ***Earth Surface Processes*** **1**, 63–70

Curtis, L.F., Courtney, F.M. and Trudgill, S.T. (1976) ***Soils in the British Isles.*** London, Longman

Curtis, L.F. and Trudgill, S.T. (1975) 'The measurement of soil moisture.' ***British Geomorphological Research Group, Technical Bulletin*** **13**, 70 pp.

Dahl, R. (1967) Post glacial micro-weathering of bedrock surfaces in the Narvik district of Norway. ***Geografiska Annaler*** **49A**, 155–166

Day, M.J. and Goudie, A.S. (1977) 'Field assessment of rock hardness using the Schmidt test hammer.' ***British Geomorphological Research Group, Technical Bulletin*** **18**, 19–29

Deju, R.A. and Bhappu, R.B. (1965) Surface properties of silicate minerals. ***New Mexico Institute of Mining and Technology, State Bureau of Mines and Technology, State Bureau of Mines and Mineral Resources, Circular*** **82**, 67–70

Douglas, I. (1968) 'Field methods of water hardness determination.' ***British Geomorphological Research Group, Technical Bulletin*** **1**, 35 pp.

Edwards, A.M.C. (1973) 'Dissolved load and tentative solute budgets of some Norfolk catchments.' ***Journal of Hydrology*** **18**, 201–217

Embleton, C. and Thornes, C.J. (1979) ***Process in Geomorphology***. London, Arnold

Emery, K.O. (1941) 'Rates of surface retreat of the cliffs based on dated inscriptions.' ***Science*** **93** (2426): 617–618

Evans, J.W. (1970) 'A method for measurement of the rate of intertidal erosion.' ***Bulletin of Marine Science*** **20**, 305–314

Finlayson, B.L. (1977) 'Measurement of soil creep at a stream bank.' ***British Geomorphological Research Group, Technical Bulletin*** **18**, 3–7

Finlayson, B.L. (1978) 'Suspended solid transport in a small experimental catchment.' ***Zeitschrift für Geomorphologie*** **22**, 192–210

Gibb, J.G. (1978) 'Rates of coastal erosion and accretion in New Zealand.' ***N.Z. Journal of marine and freshwater Research*** **12**, 429–456.

Goudie, A.S. (Ed.) (1981) ***Geomorphological techniques***. London, Allen and Unwin.

Goudie, A.S. (1977) 'Sodium sulphate weathering and the disintegration of Mohenjo-Daro, Pakistan.' ***Earth Surface Processes*** **2**, 75–86

Goudie, A.S., Cooke, R.U. and Evans, R. (1970) 'Experimental investigation of rock weathering by salts.' ***Area*** **4**, 42–48

Gregory, K.J. and Walling, D.E. (1973) ***Drainage basin form and process***. London, Arnold

Gregory, S. (1963) ***Statistical methods and the geographer***. London, Longman

Griggs, D.T. (1936) 'The factor of fatigue in rock weathering.' ***Journal of Geology*** **44**, 781–796

Haigh, M.J. (1977) 'The use of erosion pins in the study of slope evaluation.' ***British Geomorphological Research Group, Technical Bulletin*** **18**, 31–49

Hale, M.E. (1967) ***The biology of lichens.*** London, Arnold

Hanwell, J.D. and Newson, M. (1973) ***Techniques in Physical Geography***. London, Macmillan

Helley, E.J. (1969) 'Field measurement of the initiation of large bed particle motion in Blue Creek, near Klamath, California.' ***United States Geological Survey, Professional Paper*** 562–G

Hem, J.D. (1970) 'Study and interpretation of the chemical characteristics of natural water.' ***United States Geological Survey, Water Supply Paper*** 1473

High, C.J. and Hanna, K.K. (1970) 'A method for the direct measurement of erosion on rock surfaces.' ***British Geomorphological Research Group, Technical Bulletin*** 5, 22 pp.

Hodgkin, E.P. (1964). 'Rate of erosion of intertidal limestone.' ***Zeitschrift für Geomorphologie*** **8**, 385–392

Hooke, J.M. (1977) 'The distribution and nature of changes in river channel patterns:

the example of Devon.' pp. 265–280 In Gregory, K.J. (Ed.) ***River Channel Changes***. London, Wiley

Hooke, J.M. (1980) 'Magnitude and distribution of rates of river bank erosion.' ***Earth Surface Processes* 5**, 143–157

Huang, W.H. and Kiang, W.C. (1972) 'Laboratory dissolution of plagioclase feldspars in water and organic acids at room temperature.' ***American Mineralogist* 57**, 1849–1859

Imeson, A.C. (1977) 'A simple field-portable rainfall simulator for difficult terrain.' ***Earth Surface Processes* 2**, 431–436

Jones, R.J. (1965) 'Aspects of the biological weathering of limestone pavements.' ***Proceedings of the Geologists' Association* 76**, 421–434

Kirk, R.M. (1975) 'Coastal changes at Kaikoura, 1942–1974, determined from air photographs.' ***New Zealand Journal of Geology and Geophysics* 18**, 787–801

Kirk, R.M. (1977) 'Rates and forms of erosion on intertidal platforms at Kaikoura peninsula, South Island, New Zealand.' ***New Zealand Journal of Geology and Geophysics* 20**, 571–613

Knapp, B.J. (1973) 'A system for the field measurement of soil water movement.' ***British Geomorphological Research Group, Technical Bulletin* 9**, 26 pp.

Kwaad, F.J.P.H. (1970) 'Experiments on the granular disintegration of granite by salt action.' ***From Field to Laboratory, Publicatie* 16.** Fysisch Geografish en Bodenkundig Laboratoire, Amsterdam

Kwaad, F.J.P.H. (1977) 'Measurements of rainsplash erosion and the formation of alluvium beneath deciduous woodland in the Luxembourg Ardennes.' ***Earth Surface Processes* 2**, 161–173

Mercado, A. and Billings, G.K. (1975) 'The kinetics of mineral dissolution in carbonate aquifers as a tool for hydrological investigations. 1. Concentration–time relationships.' ***Journal of Hydrology*** **24**, 303–331

Morgan, R.P.C. (1978) 'Field studies of rainsplash erosion.' ***Earth Surface Processes*** **3**, 295–299

Newson, M.D. (1971) 'A model of subterranean limestone erosion in the British Isles based on hydrology.' ***Transactions, Institute of British Geographers*** **54**, 55–70

Newson, M.D. (1980) 'The geomorphological effectiveness of floods – a contribution stimulated by two recent events in mid-Wales.' ***Earth Surface Processes*** **5**, 1–16

Ollier, C.D. (1969) ***Weathering***. London, Longman

Pitty, A.F. (1971) ***An introduction to geomorphology***. London, Methuen

Potts, A.S. (1970) 'Frost action in rocks: some experimental data.' ***Transactions of the Institute of British Geographers*** **49**, 109–124

Rice, R.J. (1977) ***Fundamentals of Geomorphology***. London, Longman

Riezebos, H.T.H. and Seyhan, E. (1977) 'Essential conditions of rainfall simulation for laboratory water erosion experiments.' ***Earth Surface Processes*** **2**, 185–190

Sharp, A., Trudgill, S.T., Crabtree, R.W., Pickles, A.M., Smith, D.I., Cooke, R.U. and Price, C. (1982) 'Measurements of surface lowering of building stone, St Paul's Cathedral, London.' ***Earth Surface Processes and Landforms*** **7**, 387–389

Smith, B.J., (1977) 'Rock temperature measurements from the Northwest Sahara and their implications for rock weathering.' ***Catena*** **4**, 41–63

Smith, D.I. and Atkinson, T.C. (1976) 'The erosion of limestones.' In Ford, T.D. and Cullingford, C.H. ***The Science of Speleology***. London, Academic Press

Smith, D.I. and Stopp, P. (1978) ***The river basin***. Cambridge, Cambridge University Press

Smith, T.R. and Dunne, T. (1977) 'Watershed geochemistry: the controls of aqueous solutions by soil materials in a small watershed.' ***Earth Surface Processes*** **2**, 421–425

Thorn, C.E. (1979) 'Bedrock freeze–thaw, weathering regime in an alpine environment, Colorado Front Range.' ***Earth Surface Processes*** **4**, 211–228

Tricker, A.S. (1979) 'The design of a portable rainfall simulator infiltrometer.' ***Journal of Hydrology*** **41**, 143–147

Trudgill, S.T. (1975) 'Measurement of erosional weight-loss of rock tablets.' ***British Geomorphological Research Group, Technical Bulletin*** **17**, 13–19

Trudgill, S.T. (1976a) 'Limestone erosion under soil.' In Panos, V. (Ed.) ***Proceedings of the 6th International Congress of Speleology,*** **2**, Ba, 409–422. Prague, Academia.

Trudgill, S.T. (1976b) 'The marine erosion of limestone, Aldabra Atoll, Indian Ocean.' ***Zeitschrift für Geomorphologie Supplementband*** **26**, 164–200

Trudgill, S.T. (1976c) 'Rock weathering and climate: quantitative and experimental aspects.' pp. 59–99 In Derbyshire, E. ***Geomorphology and Climate***. London, Wiley.

Trudgill, S.T. (1979) 'Chemical polish of limestone and interactions between calcium and organic matter in peat drainage water.' ***Transactions of the British Cave Research Association*** **6**, 30–35

Trudgill, S.T., High, C.J. and Hanna, K.K. (1981) 'Improvements to the micro-erosion meter (MEM).' ***British Geomorphological Research Group, Technical Bulletin*** **29**, 3–17.

Van Zon, H. (1979) ***Litter transport as a geomorphic process.*** Publicaties van het Fysisch-Geografisch en Bodem Kundig Laboratonium van de Universiteit van Amsterdam, 24

Vear, A. and Curtis, C.D. (1981) 'A quantitative evaluation of pyrite weathering.' ***Earth Surface Processes and Landforms*** **6**, 191–198

Walling, D.E. and Webb, B.W. (1978) 'Mapping solute loadings in an area of Devon, England.' ***Earth Surface Processes*** **3**, 85–99

Waylen, M.J. (1979) 'Chemical weathering in a drainage basin underlain by Old Red Sandstone.' ***Earth Surface Processes*** **4**, 167–178

Welton, J.S. and Ladle, M. (1979) 'Two sediment trap designs for use in small rivers and streams.' ***Limnology and Oceanography*** **24**, 588–592

White, S.E. (1976) 'Is frost action really only hydration shattering? A review.' ***Arctic and Alpine Research*** **8**, 1–6

Williams, R.J.B. (1970) 'The chemical composition of water from land drains at Saxmundham and Woburn, and the influence of rainfall upon nutrient losses.' ***Report, Rothamsted Experimental Station, 1970*** **2**, 36–67

Wiman, S. (1963) 'A preliminary study of experimental frost shattering.' ***Geografiska Annaler*** **45**, 113–121

Yatsu, E., Dahms, T.A., Falconer, A., Ward, A.J. and Wolfe, J.S. (Eds) (1971) ***Research methods in geomorphology***. Science Research Associates (Canada) Ltd., 44 Prince Andrew Place, Von Mills, Ontario, Canada

Young, A. (1974) 'The rate of slope retreat.' In Brown, E.H. and Waters, R.S. ***Progress in Geomorphology***. Institute of British Geographers, Special Publication No. 7, 65–78

FURTHER READING

For full details, see Reference List

Techniques – Goudie (1981)
British Geomorphological Research Group, Technical Bulletins (*see* Appendix 1)
Hanwell and Newson (1973)

Rivers – Gregory and Walling (1973)

Water chemistry – Smith and Stopp (1978)
Hem (1970)

Chemical analysis – Allen (1967)

Process geomorphology – Embleton and Thornes (1979)
Yatsu *et al.* (1971)
Rice (1977)
Pitty (1971)
Ollier (1969)

INDEX

Citations in bold type are the main references in a group